Engineering Crop Plants for Industrial End Uses

Related titles published by Portland Press Ltd

Eicosanoids and Related Compounds in Plants and Animals
edited by A.F. Rowley, H. Kühn and T. Schewe
1998 ISBN 1 85578 108 5

Essays in Biochemistry volume 32: Cell Signalling
edited by D.J. Bowles
1997 ISBN 1 85578 071 2

Hemicellulose and Hemicellulases
edited by M.P. Coughlan and G.P. Hazlewood
1993 ISBN 1 85578 036 4

Molecular Botany: Signals and the Environment
edited by D.J. Bowles, P.M. Gilmartin, J.P. Knox and G.G. Lunt
1994 ISBN 1 85578 050 X

Plant Cell Division
edited by D. Francis, D. Dudits and D. Inzé
1997 ISBN 1 85578 089 5

Engineering Crop Plants for Industrial End Uses

Proceedings of the Symposium of the Industrial Biochemistry and Biotechnology Group of the Biochemical Society held at IACR-Long Ashton Research Station, Long Ashton, in September 1996

Editors

Peter R. Shewry
Johnathan A. Napier
Paul J. Davis

PORTLAND PRESS

Published in the United Kingdom by Portland Press Ltd,
59 Portland Place, London W1N 3AJ, U.K.
Tel: (+44) 171 580 5530; e-mail: edit@portlandpress.co.uk

In North America orders should be sent to Princeton University Press,
41 William Street, Princeton, New Jersey 08540, U.S.A.

ISBN 1 85578 113 1 ISSN 0966-4068

British Library Cataloguing in Publication Data
A catalogue record for this book is available from the British Library

Typeset by Portland Press Ltd
Printed in Great Britain by Information Press Ltd.

Contents

Preface

Agricultural production in Europe has been a huge success story.

Improvements in varieties (both plant and animal), disease control and protection from pests and diseases have allowed Europe to move from being a large importer of basic foodstuffs to being a significant presence in the world export market.

However, this success has its price, with the storage of surplus products (the well-known grain mountains and milk lakes) and subsidies being required to enable E.U. production to be exported to the lower-priced world market. Consumers have also begun to question the environmental and animal welfare aspects of current production systems.

As a result of these pressures, and realizing that in future agricultural commodity production is likely to get less support from the taxpayer, farming is looking to its centuries old role of producing fuel and fibre as well as food.

New uses are already being found for existing crops (e.g. bioethanol from oil crops) and 'new' crops are being developed (e.g. willow/poplar for coppicing for energy production, and evening primrose for pharmaceutical uses). Genetic engineering will enable farmers to produce a whole generation of new crops for specialist needs, the so-called 'designer crops', including raw materials for the chemical and pharmaceutical industries. Research into production techniques, increasing the yield per hectare, and into improved processing methods will also play vital roles in developing the supply of and demand for these new products of the agricultural industry.

However, production, and therefore the underpinning research, must be demand led. This means that all of us concerned in the development and production of both food and non-food crops, scientists, researchers, farmers, processors and retailers, must discuss and debate the issues with the customer to ensure confidence both in the product and in its methods of production. New techniques such as genetic modification, which will be essential if we are to develop new products and markets, will need to be explained to and accepted by our customers. That is why I am pleased to provide a preface to this volume, which records the proceedings of a symposium that brought together academics and industrial scientists to debate the opportunities and problems in developing new industrial crops by genetic engineering.

A.W.D. Pexton
Deputy President
National Farmers' Union

Contributors

S.A.M. Arif
Biotechnology and Strategic Research Division, Rubber Research Institute of Malaysia, RRIM Experiment Station, 47000 Sungei Buloh, Selangor D.E., Malaysia

P. Arokiaraj
Biotechnology and Strategic Research Division, Rubber Research Institute of Malaysia, RRIM Experiment Station, 47000 Sungei Buloh, Selangor D.E., Malaysia.

T.Ya. Bogracheva
John Innes Centre, Norwich Research Park, Norwich NR4 7UH, U.K.

S.-B. Choi
Institute of Biological Chemistry, Washington State University, P.O. Box 646340, Pullman, WA 99164-6340, U.S.A.

R. Croteau
Institute of Biological Chemistry, Washington State University, Pullman, WA 99164-6340, U.S.A.

P.J. Davis
Unilever Research, Colworth Laboratory, Colworth House, Sharnbrook, Bedford MK44 1LQ, U.K.

K.M. Elborough
Biological Sciences Department, University of Durham, South Road, Durham DH1 3LE, U.K.

E. Flipse
Scottish Crop Research Institute, Invergowrie, Dundee DD2 5DA, Scotland, U.K.

T.W. Greene
Institute of Biological Chemistry, Washington State University, P.O. Box 646340, Pullman, WA 99164-6340, U.S.A.

C. Halpin
Department of Biological Sciences, University of Dundee, Dundee DD1 4HN, U.K.

S. Hamzah
Biotechnology and Strategic Research Division, Rubber Research Institute of Malaysia, RRIM Experiment Station, 47000 Sungei Buloh, Selangor D.E., Malaysia

S.Z. Hanley
Biological Sciences Department, University of Durham, South Road, Durham DH1 3LE, U.K.

C.L. Hedley
John Innes Centre, Norwich Research Park, Norwich NR4 7UH, U.K.

S. Hughes
School of Biological Sciences, University of Exeter, Exeter EX4 4QG, U.K.

H. Ito
Institute of Biological Chemistry, Washington State University, P.O. Box 646340, Pullman, WA 99164-6340, U.S.A.

H. Jaafar
Biotechnology and Strategic Research Division, Rubber Research Institute of Malaysia, RRIM Experiment Station, 47000 Sungei Buloh, Selangor D.E., Malaysia

E. Jacobsen
Graduate School of Experimental Plant Sciences, Department of Plant Breeding, Agricultural University Wageningen, P.O. Box 386, 6700 AJ Wageningen, The Netherlands

M.E. John
Agracetus, a unit of Monsanto, 8520 University Green, Middleton, WI 53562, U.S.A.

J.E Johnson
Department of Molecular Biology, The Scripps Research Institute, 10666 North Torrey Pines Road, La Jolla, CA 92037, U.S.A.

H. Jones
Division of Biosciences, University of Hertfordshire, Hatfield, Herts AL10 9AB, U.K.

T. Katsube
Shimane Women's College, Matsue, Shimane 690, Japan.

H. Kavakli
Institute of Biological Chemistry, Washington State University, P.O. Box 646340, Pullman, WA 99164-6340, U.S.A.

C. Kjøller
Bioraf Denmark Foundation, Research Centre, P.O. Box 35, DK-3720 Aakirkeby, Denmark

M.E. Knight
Zeneca Plant Sciences, Jealott's Hill Research Station, Bracknell, Berks RG12 6EY, U.K.

A.J. Kortstee
Graduate School of Experimental Plant Sciences, Department of Plant Breeding, Agricultural University Wageningen, P.O. Box 386, 6700 AJ Wageningen, The Netherlands

J.C. Kridl
Calgene, Inc. 1920 Fifth Street, Davis, CA 95616, U.S.A.

A.G.J. Kuipers
Graduate School of Experimental Plant Sciences, Department of Plant Breeding, Agricultural University Wageningen, P.O. Box 386, 6700 AJ Wageningen, The Netherlands

M.J. Laughlin
Institute of Biological Chemistry, Washington State University, P.O. Box 646340, Pullman, WA 99164-6340, U.S.A.

T. Lin
Department of Molecular Biology, The Scripps Research Institute, 10666 North Torrey Pines Road, La Jolla, CA 92037, U.S.A.

G.P. Lomonossoff
Department of Virus Research, John Innes Centre, Colney Lane, Norwich NR4 7UH, U.K.

N. Maruyama
Research Institute for Food Science, Kyoto University, Uji, Kyoto 611, Japan

D. McCaskill
Institute of Biological Chemistry, Washington State University, Pullman, WA 99164-6340, U.S.A.

B.J. Miflin
Institute of Arable Crop Research, IACR-Rothamsted, Harpenden, Herts AL5 2JQ, U.K.

M.M. Moloney
Department of Biological Sciences, The University of Calgary, 2500 University Drive N.W., Calgary, Alberta T2N 1N4, Canada

J.A. Napier
IACR-Long Ashton Research Station, Department of Agricultural Sciences, University of Bristol, Long Ashton, Bristol BS18 9AF, U.K.

A. O'Connell
Zeneca Plant Sciences, Jealott's Hill Research Station, Bracknell, Berks RG12 6EY, U.K.

T.W. Okita
Institute of Biological Chemistry, Washington State University, P.O. Box 646340, Pullman, WA 99164-6340, U.S.A.

D.W. Ow
Plant Gene Expression Center, USDA-Agricultural Research Service, 800 Buchanan St., Albany, CA 94710, U.S.A.

A.W.D. Pexton
Deputy President, National Farmers Union, Agriculture House, 164 Shaftesbury Avenue, London WC2H 8HL, U.K.

C. Porta
Department of Virus Research, John Innes Centre, Colney Lane, Norwich NR4 7UH, U.K.

P. Salamone
Institute of Biological Chemistry, Washington State University, P.O. Box 646340, Pullman, WA 99164-6340, U.S.A.

S.L. Sanda
Lipid Molecular Biology Group, Department of Biological Sciences, University of Durham, South Road, Durham DH1 3LE, U.K.

W. Schuch
Zeneca Plant Sciences, Jealott's Hill Research Station, Bracknell, Berks RG12 6EY, U.K.

P.R. Shewry
IACR-Long Ashton Research Station, Department of Agricultural Sciences, University of Bristol, Long Ashton, Bristol BS18 9AF, U.K.

C.S. Sidebottom
Unilever Research, Colworth Laboratory, Colworth House, Sharnbrook, Bedford MK44 1LQ, U.K.

A.R. Slabas
Biological Sciences Department, University of Durham, South Road, Durham DH1 3LE, U.K.

M.A. Smith
School of Biological Sciences, University of Bristol, Bristol BS8 1UG, U.K.

V.E. Spall
Department of Virus Research, John Innes Centre, Colney Lane, Norwich NR4 7UH, U.K.

K. Stephens
Institute of Biological Chemistry, Washington State University, P.O. Box 646340, Pullman, WA 99164-6340, U.S.A.

A.K. Stobart
School of Biological Sciences, University of Bristol, Bristol BS8 1UG, U.K.

F. Takaiwa
National Institute of Agrobiological Resources, Tsukuba, Ibaraki 305, Japan

K.M. Taylor
Department of Virus Research, John Innes Centre, Colney Lane, Norwich NR4 7UH, U.K.

S. Utsumi
Research Institute for Food Science, Kyoto University, Uji, Kyoto 611, Japan

C.P.E. van der Logt
Unilever Research, Colworth Laboratory, Colworth House, Sharnbrook, Bedford MK44 1LQ, U.K.

R.G.F. Visser
Graduate School of Experimental Plant Sciences, Department of Plant Breeding, Agricultural University Wageningen, P.O. Box 386, 6700 AJ Wageningen, The Netherlands

T.L. Wang
John Innes Centre, Norwich Research Park, Norwich NR4 7UH, U.K.

A.J. White
Biological Sciences Department, University of Durham, South Road, Durham DH1 3LE, U.K.

R. Woodbury
Institute of Biological Chemistry, Washington State University, P.O. Box 646340, Pullman, WA 99164-6340, U.S.A.

H.Y. Yeang
Biotechnology and Strategic Research Division, Rubber Research Institute of Malaysia, RRIM Experiment Station, 47000 Sungei Buloh, Selangor D.E., Malaysia.

Abbreviations

ABC	ATP-binding cassette
ACP	Acyl carrier protein
AGP	ADPglucose pyrophosphorylase
1AT	Glycerol-3-phosphate-1-acyltransferase
2AT	1-Acyl-glycerol-3-phosphate-acyltransferase
CAD	Cinnamyl alcohol dehydrogenase
CaMV	Cauliflower mosaic virus
CCoAOMT	Caffeoyl-CoA O-methyltransferase
CCR	Cinnamoyl-CoA reductase
CCT	CTP:phosphocholine cytidyltransferase
CDP	Cytidine diphosphate
CDR	Complementarity determining region
C3H	Cinnamate 3-hydroxylase
C4H	Cinnamate 4-hydroxylase
COMT	Caffeate/5-hydroxyferulate O-methyltransferase
CPMV	Cowpea mosaic virus
CPT	CDP-choline:diacylglycerol cholinephosphotransferase
CTP	Cytidine triphosphate
DMAPP	Dimethylalkyl diphosphate
DPA	Days post-anthesis
DSC	Differential scanning calorimetry
ER	Endoplasmic reticulum
F5H	Ferulate 5-hydroxylase
FPP	Farnesyl diphosphate
Fruc 1,6-P_2	Fructose 1,6-diphosphate
GAP	Glyceraldehyde 3-phosphate
GBSS	Granule-bound starch synthase
GGPP	Geranylgeranyl diphosphate
Glc 1-P	Glucose 1-phosphate
GPP	Geranyl diphosphate
GUS	β-Glucuronidase
HMGR	3-Hydroxy-3-methylglutaryl-CoA reductase
HRV-14	Human rhinovirus 14
IPP	Isopentenyl diphosphate
LPAAT	Lysophosphatidic acid acyltransferase
LS	Large subunit
mAb	Monoclonal antibody
MCAT	Malonyl-CoA:ACP transacylase
MVA	Mevalonic acid
PAL	Phenylalanine ammonia-lyase
PC	Phosphatidylcholine
PCn	Phytochelatin
PE	Phosphatidylethanolamine
PEP	Phosphoenolpyruvate
3-PGA	3-Phosphoglycerate
PHA	Polyhydroxyalkanoates
PHB	Polyhydroxybutyrate

PHD	Poly(3-hydroxydecanoate)
PHV	Polyhydroxyvalerate
PS	Phosphatidylserine
PUFA	Polyunsaturated fatty acid
RRIM	Rubber Research Institute of Malaysia
sc	Single chain
SRP	Signal recognition particle
SS	Small subunit
SSS	Soluble starch synthase
TAG	Triacylglycerol
TC	Thermal conductivity
TGA	Thermogravimetric analysis
TMV	Tobacco mosaic virus
T_o	Temperature of onset of gelatinization
WASP	Wiskott–Aldrich syndrome protein

Crops for the 21st century

Ben Miflin
Institute of Arable Crops Research, IACR-Rothamsted, Harpenden,
Herts AL5 2JQ, U.K.

The title should not be taken to imply that I have the answers or feel moved to make any predictions. If I did so I would wish to preface them with the thought that any prediction I might make will be wrong except for this one. Rather, I would like to take this opportunity to outline some of the factors that we have to consider in developing new crop technologies for the uncertain future of the 21st century. In doing this I will draw on discussions I have been having with a number of colleagues on the Technology Foresight Panel for Agriculture, Horticulture and Forestry. I would like to pay tribute to their inputs but to absolve them from all blame for my interpretation.

In predicting what might happen in the next century, it is necessary to look at what might be the market forces that would stimulate certain technologies and alter the sort of crops we might grow, as well as to consider the technologies that are being developed and which might push certain changes into existence. I will also discuss some of the key drivers that might operate in the future and the different sorts of scenarios that may result from these.

Historical background

Before starting to look forward, it may be interesting to look back over the changes that have occurred in the 20th century. The most dramatic have probably related to mechanization and the reduction of the labour demand of agriculture in the western world. This has been accompanied by the development of a range of technologies that have tremendously increased the yields of our major crops. The Broadbalk experiment at Rothamsted has recorded the changes in agronomic practice and cereal breeding over the last 150 years. Although Lawes, the founder of Rothamsted, had already demonstrated the effects of fertilizers, the majority of the potential that could be achieved by their use had not been realized. The introduction of improved varieties and the use of herbicides and fungicides have allowed the yield on the well-fertilized plots to increase by 3-fold, most of which has been achieved since [1a].

The range of crops in the U.K. has also changed considerably for various reasons. Some of these changes were induced by catastrophic happenings (e.g. the Second World War), others by political decisions, especially those of the Common Agricultural Policy of the European Union. These changes involve the virtual disappearance of oats and certain root crops and the appearance of new crops, such as sugar beet in the 1920s and oil-seed rape in the 1970s. There have also been dramatic changes in the relative importance of wheat and barley. Such changes may well occur in the future, powered by a number of different 'drivers'. It is probably helpful in predicting the future to identify some of those drivers.

Key drivers

Politics and economics

The last decade has encompassed a number of key international agreements that will have dramatic impacts on world agriculture as their repercussions work through the political systems. The ongoing negotiations on World Trade (GATT and its successors) have reduced supports by different governments for a range of crops. If these are followed by further relaxation of supports, then prices of major crops will move towards the world market price and farmers' decisions will be based more on these in relation to their costs than on support prices. During this period the E.U. has extended the membership of the community and plans to extend it further to include members of the former Communist bloc, which itself has disintegrated since 1989. Countries in this bloc, such as Hungary and the Ukraine, have large areas of good agricultural land which, for various reasons, have been under-producing. Inclusion of Hungary and other E. European countries into the E.U. will stimulate their agriculture and make them major competitors to W. European farmers. The Ukraine is unlikely to join the E.U. but could still make considerable strides in arable production given sufficient investment and a suitable political system. All of these issues are capable of bringing about considerable change in the production of arable crops within Europe.

Political support of agriculture in Europe has owed much to the effect of disruption of food supplies during World War II and the strength of the political lobby of farmers. However, mechanization has drastically reduced the number of voters concerned with agriculture and thus the power of the lobby, although in some countries (e.g. France) the public relations strengths of the farmers has kept the lobby strong. The perception of food shortages has been replaced by that of surpluses. This perception is not totally borne out by facts, particularly the fact that the world's reserves of wheat have been below the Food and Agriculture Organization (of the UN) defined minimum for at least five years. If the perception swings back due to a rapid rise in the price of bread or some shortages in the shops, the political pressure on governments to 'do something' will be considerable and could bring about drastic changes. The volatile nature and the power of European public opinion to cause an abrupt change in farming and the economics of the food chain have recently been demonstrated by the BSE crisis. A political change in perceptions, with or without a factual basis, can drive agriculture in different directions.

Politics and regulations

The regulatory environment of Western countries (e.g. the E.U. and U.S.A.) is a major factor in determining which agrochemical products come to market. Not only do the regulators make decisions on specific products but the cost of producing a package of information to satisfy the requirements is a major factor in companies' decisions whether to develop a product for a given market. Thus, areas of many crops are too small to justify the development and registration of crop protection products to suit their needs. Similarly, products need to be registered over several countries to ensure the market is large enough to provide a return on investment.

The same sorts of considerations will affect the development of products from recombinant DNA technology. Currently, the public acceptance of biotechnology is under question in Europe and this has influenced the E.U. to develop strict regulations on the experimental and commercial release of transgenic crops. In the early days, these regulations were similar on both sides of the Atlantic but in 1992 the U.S.A. reduced the regulations on several crops with which there had been considerable experience. Consequently, some 90% of field trials in the U.S.A. now take place under a simple notification procedure. Procedures for approving the commercial release of transgenic crops in the U.S.A. have also been streamlined and currently take less than a year to complete. In contrast, E.U. regulations for commercial release involve a complex procedure involving approval by all countries that currently takes about two years. As a result, there is a large acreage of transgenic crops being grown this year in N. America and virtually none in the E.U. This differential approach may lead to disputes between the U.S.A. and E.U., particularly since the crops include maize and soybeans that are imported into the E.U. The way in which the situation develops will have consequences in terms of the speed with which new crop technology is developed for the benefit of European farmers.

As yet (September 1996), common regulations for the use of products from genetically modified plants in the food and feed chains have not been agreed by the E.U. Individual countries are, therefore, free to make their own decisions and the U.K. has approved the use of vegetable oils from canola (oil-seed rape) and soybean and of paste from transgenic tomatoes. The latter is now on sale in two supermarket chains. There is considerable debate as to whether the labelling of products should be mandatory but the companies have decided to label the tomato paste voluntarily. This differential development of regulations may lead to products from transgenic crops being on sale in European shops before it is possible to grow the crop in the fields of the E.U.

The Rio Summit in 1992 led to the signing of a Biodiversity Convention by most countries except the U.S.A. The consequences of this convention are currently being worked out and are likely to govern the transfer of genetic resources and plant genes between countries. There is also considerable debate about the protection of intellectual property through patents, plant breeder's rights and other methods. Again there are different rules and regulations in the U.S.A. from elsewhere. The availability of protection will affect both the costs of, and the returns from, investing in crop biotechnology.

The future development of regulations in the U.K. will depend on the way in which politicians perceive the public acceptance of recombinant crops. In this the pressure groups as well as the general public will play a part. Although there are a large number of activists, the evidence from the sales of the two labelled products of recombinant DNA currently available in British supermarkets (i.e. cheese made from recombinant chymosin and tomato paste) do not indicate a great degree of consumer concern. Similarly, the recombinant fresh tomatoes on sale in the U.S.A. appear to have been well accepted.

Market economics

Which crops are grown will depend on their profitability to the farmer; in part this depends on the degree of protection they receive but eventually on the demand for them. In broad terms, certain crops are likely to be favoured by changes in eating patterns as, for example, in the much wider demand for wheat relative to other cereals. Which crops receive technology investment will also depend on their market size and profitability. The costs of registering new pesticides are such that this is only justified if the compounds are active against problems in major crops. This is likely to bring about a circle in which the most widely grown crops get the best agrochemical technology, which makes them more profitable, which makes them more successful etc. The converse is true for minor crops.

Technologies that are sold through the seed (e.g. plant breeding) will be subject to the market situation for seed, which is dramatically different to that for agrochemicals. In general, the returns on inbred seed are much less than for hybrid crops, chiefly because the farmer can profitably save and re-use the seed from an inbred whereas he cannot from a hybrid. This is likely to affect the ability of investors in advanced technologies like plant transformation or molecular markers to get a return from their use for inbreds.

Environmental change

Climate change appears now to be generally accepted. This is likely to be manifested in both a steady change, e.g. increases in average temperature, and in increased variability in temperature. Whilst undoubtedly the changes in average temperature and rainfall will impact on crops, the most recent higher resolution models suggest that the effects in the U.K. will not be great [1]. However, when the effects of increased variability in climate are tested on existing models of wheat growth the results predict that such variability will have considerable effects on yields. Tolerance to such variability will be hard to select for by current approaches to plant breeding. Climate change, if it occurs, may bring more difficulties to continental European farmers whereas those in the U.K. are likely to be more buffered against the effects.

Environmental change is also likely to affect the distribution of pests and diseases. For example, long-term monitoring of insect pests in the Rothamsted Insect Survey predicts that the importance of aphids and the virus diseases that they transmit as vectors will increase considerably as the temperature rises [2].

Future scenarios

Based upon the above, and a little imagination, it is possible to come up with a range of different scenarios for the future. It is important to try to encompass a range of possibilities instead of trying to portray the most likely one because it is sure that that one will be wrong. The aim of the scenarios is to stimulate thought of possible alternative courses of action that should enable us to meet the changes that may occur. Three possible scenarios are as follows.

Surpluses

Europe has been in surplus for the production of almost all of its required arable crop produce for some years; only in protein and oil crops is there some form of deficit within the community. Common Agricultural Policy support of farmers through this time of surplus has cost the E.U. a large amount of money and is currently under reform. This scenario assumes that new technologies will boost yields more than expected, that the population growth is at the low end of predictions and that the agriculture in E. Europe and S. America responds positively and dramatically in increasing production. Reasons for such a scenario are given in more detail in a recent book by Dyson [3] in which he concludes that in all probability the people of the world will be somewhat better fed in the year 2020 than they are today.

Environmental

A situation of local surpluses and high Gross National Product in Europe, and an absence of any pressures on production from outside Europe, coupled with the increased concern with environmental and safety issues, could lead to a situation in which these issues were the primary drivers for agricultural production. In such a scenario, the influence of European-wide regulations and a swing back from the de-regulatory philosophy of recent years, would lead to strict limits on the freedom that producers would have in their production methods. The high incomes of the population would lead to demands for more quality and variety in the foods available.

Shortages

In this scenario the current fluctuations in the price of wheat, which increased in 1995 as stocks decreased and fell in 1996 as more land came back into cultivation in the U.S.A. and Europe, would continue. The growth in the world population and the rise in living standards in Asia, particularly China, would greatly increase the demand for cereals and local production would not be able to respond quickly enough. More land would be brought back into production in Europe and N. and S. America and would allow us to cope for a time but eventually the demands would exhaust capacity and a state of permanent shortage would develop. In this scenario some of the technical developments that are currently foreseen would not work out for technical or infrastructure reasons, possibly aided by more pests and diseases as production intensity went up. Climate change would occur with some dramatic consequences. The argument has been advanced by the pessimistic school of futurologists, who are well represented by Lester Brown [4]. In such a case, the overall political philosophy would be non-interventionist and world trade would continue to open up. Food shortages would override people's concerns with the environment and regulations.

Technical responses

These scenarios would pull technology in different ways. In the case of surpluses, the lowest-cost producers would provide the bulk of production and this might be largely outside the U.K. and even W. Europe. Producers would want technologies that gave them the lowest cost of production (e.g. in-built resistance to pests and diseases) and the most effective use of inputs (support to make the right decisions). Higher-cost producers would opt for niche markets and added value from the food chain. They would tend to try to cash in on the perceived values of 'organic' and 'natural' produce. They would also be looking for crops with specialist traits and qualities (e.g. special oils) and experiment with a wider range of vegetables. A proportion of current U.K. arable land may go out of production and be used for its 'amenity' value.

The regulatory scenario would only allow clean, 'risk-free' and labelled products of biotechnology on to the market. There would be an upsurge in demand for identity-preserved production and traceability. This would put demands for the development of diagnostics and decision/precision agriculture. It might, for example, lead to pesticide use being 'on prescription' from plant doctors. There would be a development of crops that had perceived or real benefits for health. The public and political perception of technologies and products would have a very strong influence on the market and changes in that perception might be frequent and rapid. Production flexibility would therefore be critical.

The shortages scenario would require staple crops grown on a wider area. There would be a premium on high yields and inputs would in the main be used prophylactically. Relaxed regulations and the demand for new technologies for crop improvement would increase the investment in research, particularly in the public sector, and the use of genetically modified organisms would increase rapidly. There would be little demand for specialist qualities and high yields would be at a premium. Decision/precision agriculture would only be used on a large scale to ensure maximum production.

In summary, the various scenarios would lead to very different cropping patterns and effects on the face of the countryside in the U.K. Agrochemical businesses would have very different futures, as would businesses based on specialist diagnostics and small-scale precision equipment, similarly the nature of the biotechnology products and the speed of their introduction might be somewhat different under the different scenarios.

Technology push

Recent research has provided a number of genetic marker technologies that are already being used by breeders to improve crop plants, particularly in hybrids such as maize where the value of the seed is high. The cost of these is likely to fall and the ease and speed of the assays markedly increase. These will greatly improve the breeding process and increase the number of traits that can be handled by breeders. In addition, in the next 5–10 years the complete sequences of the genomes of two plant species (*Arabidopsis* and rice) are likely to be produced. The potential for understanding the genetic controls on many important crop traits will increase considerably and breeders will be much better able to handle the variation present in their crops.

However, not all the desirable genes that a plant breeder may want to use are present within the crop species that he is improving. The solution to this, if a suitable gene exists elsewhere or can be manufactured, is to introduce it via transformation. The pace of transformation technology has been very rapid. The first examples of transgenic plants inheriting and expressing introduced traits were only reported in 1983 but now, some 13 years later, commercial crops are in the ground in over a million hectares in the U.S.A. and Canada. The crops include soybeans, maize and cotton, which are not major contributors to U.K. agriculture, and oil-seed rape which is. The traits are mainly related to crop protection targets, including insect resistance and herbicide tolerance. However, in the pipeline are a large number of traits that could change the nature of the products of crops. The most developed are those that change the ripening of tomatoes, where two products are already on the market. Shortly behind these are modifications to the fatty acid composition of oil-seed rape. A summary of some of these advances is given in the proceedings of a recent conference [5] as well as in papers elsewhere in this volume. In addition, the transformation of wheat and barley has been achieved and some progress is being made in introducing useful traits into these species [6]. Providing that investment continues we may expect to see an increasing number of transgenic crops reaching the market (but see the section on market economics above). How soon transgenic cereals play an important part in U.K. agriculture is an open question, in any event it will be some time after transgenic crops have made a major impact in N. America.

The other area where technology push may change the nature of arable crops in the U.K. is in their management through the use of decision support systems. The explosion in information technology will allow information to be gathered and models to be developed to predict the importance of such information. Such predictions may be made to assist farmers to make decisions regarding the growing of their crops. For example, the power of molecular biology to recognize organisms very specifically and to determine some of their properties (e.g. resistance to agrochemicals) will allow the development of diagnostics to alert farmers to potential dangers in their fields. Allied to this power of recognition is the development of technologies for the precise delivery of agricultural inputs. In the environmental scenario outlined above, it may be expected that these decision/precision technologies will be widely used to minimize the use of inputs and to ensure they are only used as needed. The speed of introduction of these technologies will be largely driven by regulations. In other scenarios, they would be used according to the cost benefit they could deliver.

Long-term challenges

The first part of this chapter has dealt with some of the possibilities that could arise given the technologies and knowledge we can foresee at present. The possibilities that do make it to the market will depend on how the economic and trade environment develops in the next decade or so. Beyond that time-frame, we may move into the later stages of the shortages scenario as the world population increases up until 2050, when it is predicted to stabilize. Whether one follows the predictions of the optimists [3] or the pessimists [4], they agree that there will be a continued need for agricultural research to ensure food supplies. What then might be some of the challenges that research needs to conquer to ensure the optimists remain optimistic up until 2050?

One certain challenge the researchers will face in the near future is to transform the information obtained in sequencing the genome into knowledge of how plants work. This will require a renewed interest in physiology and the development of molecular physiology as a growing discipline. The first findings will be the easy ones in which visible traits that are relatively independent of environment will be understood. However, crop performance in the field is dependent on the often subtle interaction of genotype with the environment; finding out how genes control plant performance under a range of different environments and in a variety of different genetic backgrounds is very much more of a challenge.

Another crucial challenge arises out of the realization that most of the gains in crop productivity have been realized through breeders manipulating the harvest index, i.e. by increasing the proportion of the total biomass accumulated during the season that is present in the harvested portion of the crop (for an excellent account of work in this area see the book by Evans [7]). Studies have shown (e.g. Riggs and co-workers [8]) that, in the barley varieties released in the U.K. over the last 100 years, the total amount of biomass accumulated during the season has hardly changed. Similar situations exist for other crops. The challenge to increase basic crop productivity in terms of total biomass is one that remains to be faced.

Another long-term challenge is the eventual decline in the amount of fossil fuel available to the world. The majority of fuel used in the world is dependent on the products of past photosynthesis stored as oil, coal and gas. These stocks are finite. As shortages in fuel drive up the prices, the use of fossil energy in crop production and in the distribution of the harvested product becomes a cause for concern and may well affect arable agriculture dramatically. The second consequence will be the use of current crops to provide energy and chemical raw-materials to replace those derived from fossil fuels. Some ideas on how this might develop are given in papers at this conference but we may expect many more such developments in the future.

Crop production is crucially dependent on water, and water shortages are already evident in many parts of the world. Whilst there are many ways in which current shortages might be overcome by improved technology and handling of water, the overall water use efficiency of many of our crops could possibly be improved. For example, our temperate cereals, which fix carbon via the C3 pathway of photosynthesis, use between 330 and 1000 g of water for each gram of CO_2 fixed. In contrast, tropical cereals such as maize and sugar cane, which use the C4 pathway of carbon fixation, only need about half this amount. A challenge to plant engineers is thus 'can the water use efficiency of major cereals like wheat and rice be improved?' In addition, climate change may well increase water and other stresses on plants so that new levels of stress resistance will be required.

Jerome K. Jerome once wrote 'you should always buy land because they are not making any more'. Estimates suggest that we are probably already using some 90% of the suitable land for agriculture. In addition, the increase in population will put pressures on the use of that land for other purposes. This will challenge agronomists and soil scientists to ensure that the biological value of the soil is maintained into the future.

All of these challenges suggest that there is every reason to increase our efforts in crop research. Even if the mid-term future follows the surpluses scenario outlined above, the total direct or indirect dependence of man on photosynthesis for his food and a substantial part of his energy requires that the productivity of crops continues to be raised by all appropriate technology. Thus, if I am forced to make a prediction for crops for the 21st century it would be that they will be more important than ever for man's continued existence.

References

1a. Johnston, A.E. (1994) in Long-Term Experiments in Agricultural and Ecological Sciences (Leigh, R.A. and Johnston, A.E., eds.), CAB International, Wallingford
1. Semenov, M.H. and Barrow, E.M. (1997) Climatic Change **35**, 397–414
2. Harrington, R. and Woiwod, I.P. (1995) Weather **50**, 200–208
3. Dyson, T. (1996) Population and Food: Global Trends and Future Prospects, Routledge, London, 231 pp.
4. Brown, L.R. (1994) in State of the World 1994 (Brown, L.R., Flavin, C. and Postel, S., eds.), Earthscan Publications Ltd, London
5. Collins, G.B. and Shepherd, R.J. (eds.) (1996) Engineering Plants for Commercial Products and Applications, Ann. N. Y. Acad. Sci. **792**, 183 pp.
6. Shewry, P.R., Tatham, A.S., Barro, F., Barcelo, P. and Lazzeri, P. (1995) Biotechnology **13**, 1185–1190
7. Evans, L.T. (1993) Crop Evolution, Adaptation and Yield, Cambridge University Press, Cambridge, U.K., 500 pp.
8. Riggs, T.J., Hanson, P.R., Start, N.D., Miles, D.M., Morgan, C.L. and Ford, M.A. (1981) J. Agric. Sci. (Camb.) **97**, 599–610

Chemicals from plants: an industry view

Steve Hughes*
Unilever Research, Colworth Laboratory, Colworth House, Sharnbrook, Bedford MK44 1LQ, U.K.

Introduction

Whatever the motive for the development of plants to provide raw materials or sophisticated products for the chemical industry, whether it be the redeployment of surplus agricultural resources, or a switch away from dependence on fossil resources, any consideration of feasibility should include a wider view of the factors that are key to success. In addition to technical feasibility these factors will inevitably include the ability of the agricultural supply chain to adapt to the changes in practice that will be required, as well as the social and political consequences of these changes. Of similar importance will be the ability to appreciate the commercial parameters governing the practical feasibility of individual technical opportunities. These will include market size (supply and demand), product value, flexibility of demand, the cumulative and relative costs of the various steps in the production chain, and the implications of by-products. Underlying all of this, from the conceptual and development phases onwards, is a requirement for information flow and a constructive dialogue between participants in the production chain.

Evaluation of the commercial opportunity

Although each opportunity ultimately must be evaluated individually on the basis of market forecasts some generalizations are possible.

First, there is likely to be an inverse relationship between product value and market size. The limits of value are set by agricultural commodity values (US$200–300 ton^{-1}) at the low end of the value scale and by pharmaceutical product values (US$1000s kg^{-1})at the upper end. There is little scope for the generation of large added value within the agricultural production chain, at these extremes, since at the commodity end there will be no margin and at the upper end no volume. Between these extremes, opportunities for margins from value added in the agricultural phase will depend upon the relative shapes of the value–volume and the demand–supply curves.

* Present address: School of Biological Sciences, University of Exeter, Exeter EX4 4QG, U.K.

Beyond this simple cost-based rationale, possibilities for the redeployment of land or other fixed resources (labour and machinery) may render the uptake of new crops or alternative commodity crops attractive to farmers provided that available margins are sufficient to offset additional supply chain costs (see below).

Agricultural productivity in terms of the amount of feedstock or intermediate produced per hectare is another important commercial parameter, although it may assume lesser importance in the light of the relative costs of other activities in the production chain (see cumulative costs below). One useful reference marker is the lower limit of production costs for the competitive microbial fermentation route to the production of chemical feed stocks and intermediates, which is of the order of US$5 kg^{-1} [1].

Flexibility of demand becomes an issue once feasibility, production and market saturation are established. Any subsequent improvement of productivity, unless linked to an increase in market size, will reduce the scale of demand for the agricultural product and thus the attractiveness of the crop to the farmer.

The relative scale of the cumulative costs in the supply chain from seed production through to product extraction and marketing is also key to determining both the attractiveness of a venture to the diverse actors involved (see production chain issues below) and the technical capabilities that are key to success. This is the phenomenon of *cost sensitivity*. For example, if the agricultural costs of supplying the raw material are a small component in the overall chain relative to the costs of isolation and purification of the product, then it will make sense to focus technology on the isolation steps rather than on designing plants with enhanced yield of the product or precursor relative to the commodity crop.

The production chain

The agro-industrial production chain as we know it has evolved for the efficient production, preservation and distribution of food. The relationship between actors therefore represent historical linkages and traditional values coupled to incrementally changing market conditions.

An outline diagram representing the sequence of actors linked by their sequential inputs and outputs is shown in Figure 1. This diagram illustrates also the parallel animal production chain, the relevance of which becomes apparent when we consider the disposal of by-products.

Under normal circumstances in open markets there exists a tension as each actor seeks to optimize its own input and output prices within the constraints of commodity fluctuations. This function reflects the general interchangeability of commodities and commodity-derived products and the multiplicity of participants in each actor group. Under these circumstances sustained relationships or partnerships are not common.

In the face of opportunities for radical changes in the supply and demand relationships in the chain there will be a requirement for cooperation and the establishment of close and inflexible linkages between participants. This will be reflected most in the requirement for *identity preservation* or raw material

Figure 1

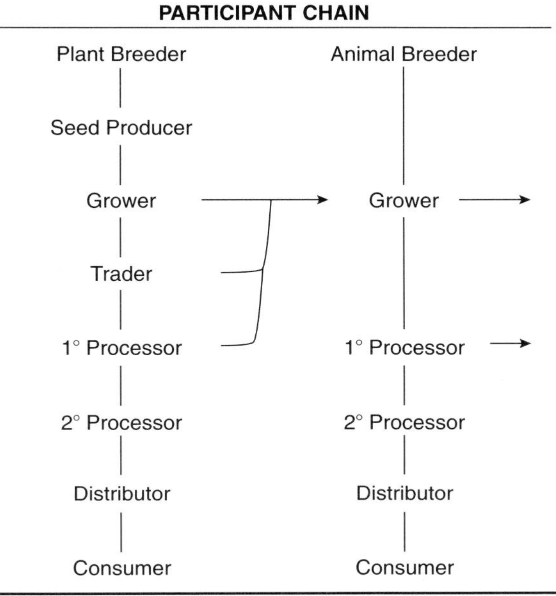

Sequence of actors in the agro-industrial production chain

segregation, as we move away from commodity sourcing. The concept of identity preservation in agricultural products is not entirely novel to the production chain. For example, it is well represented at the quality end of the champagne trade where provenance and quality of raw material is key to product authenticity. It is perhaps to such models that we need to look for success in developing the chain for the production of chemicals.

This raises the key issue of leadership, i.e. leadership in terms of setting targets, taking the lead in establishing the chain, and setting priorities for the supporting research and development phase. In the 'identity preserved chain' leadership generally comes from the end user or the last actor in the chain. However, in the case of engineered plants, leadership will also come from the top of the chain, generated through the vision and opportunism of the plant breeder and plant scientist. Is the agricultural sector prepared to surrender leadership to other and potentially remote actors in its quest for new outlets for spare productive resources? The challenge for this sector will be to stay closely engaged with the other potential actors as well as with the technical developments that will govern feasibility.

Given that leadership is likely to come from the ends of the chain an obligation of communication is placed on those actors and their supporting technologists. Effective communication of the basis for both the technical and the commercial opportunities from an early stage will be an essential component in building confidence among potential partners in the chain.

On a day to day basis, as the production chain is established and the productive cycle implemented, a new cadre of individual leaders will be required.

These will be multi-skilled managers with the ability to understand the diversity of functions from technology through production to marketing.

'Identity preservation' will bring its own additional costs in the form of a premium on the commodity crop value. This brings us back to the wider issue of the relative added value (costs) accumulated at each step in the chain as a function of the end value of the product. This has impact not only on priorities for technical development as discussed above, but also on the evaluation of the commercial opportunity. Heavy costs can be expected in the post-agricultural phase in the extraction and purification of the product, especially at the extremes of Good Manufacturing Practice required in the health sector. The implications of the disposal of bulk by-products should also be mentioned at this point. It is possible that by-products of crops designed for the production of industrial feed stocks will not be permitted to enter the normal by-product disposal route through animal feed. In this case alternative disposal provisions will be required and these could add significantly to the overall costs of production. Such cumulative costs will have the effect of diminishing the relative significance and the leverage of the agricultural production sector.

Sociopolitical aspects

We are accustomed to thinking of agricultural production in the context of food security and all of the issues of locally protected markets, protected employment, and commodity price support which this implies. The current trend is for international trade agreements (GATT, NAFTA) to shift this position towards unbuffered global markets and prices for commodities. This will inevitably complicate the supply-chain considerations with respect to large-scale crop production for chemical feed stocks, especially where the crop value is close to commodity values. This may deny opportunities for local economic microclimates within which to generate commercial feasibility.

The advent of set-aside arrangements as a mechanism for dealing with potential overproduction has raised demand in the farming community for alternative non-food crops for the utilization of set-aside land, labour and machinery. For as long as these support arrangements hold they will act as a stimulus to the development of alternative crops and crop uses.

The possibility of developing crops to produce raw materials normally produced by other crops in other climates and other countries raises the important issue of import substitution. This is especially significant where it could potentially reduce the market for the products of developing countries that are deeply committed to and dependent upon export of primary products. There is nothing unique in agricultural technology in this regard, all technological developments have the potential to alter trading patterns. However, opponents of the use of new technologies in agriculture frequently articulate this issue as a serious downside of agricultural change so it needs to be considered with sensitivity within the social dimension.

Reliance on finite fossil resources for energy and feed stocks, is in the longer term, non-sustainable. A switch to renewable resources such as those coming from agriculture, provided that it does not interfere with sustainability of food production, is potentially beneficial in terms of reducing shorter-term pressures on non-renewables and extending their utility. Moreover, as we shall see, biologically derived materials are generally biodegradable and in these days where product life-cycle analysis is becoming *de rigueur*, they bring the added benefit of lower environmental impact.

Practical requirements for success

Reverting now from the needs and requirements for success in individual ventures, let us examine the more general and methodological needs of the field as a whole.

Information is an obvious necessity but beyond that lies the ability to rapidly access information sources (in particular electronic databases) and to constructively link diverse sources together. At the moment diverse databases do exist for patterns of use of natural products, natural product chemistry, plant metabolism and plant genetics. However, these have evolved separately and are not necessarily structured compatibly. As a consequence we lack the means to link or co-interrogate them.

Recently an initiative named ACTIN (Alternative Crops Technology Interaction Network) has been set up under the auspices of the Biotechnology and Biological Sciences Research Council and industry to confront this issue. There is clearly a demand for the most advanced tools of Information Technology (commonly referred to as Bioinformatics in this scenario).

At the technical level, with respect to the engineering of plants for specialized productivity, we require an enhanced understanding of metabolic control both in terms of the molecular and genetic mechanisms and the consequences of remodelling of metabolic pathways. Stated thus this is a fairly nebulous requirement, akin to saying that we require good science. However, in the course of meetings such as the one reported in this volume we expect to see a refinement of this target to specific sectors of specific metabolic pathways and biosynthetic processes, as well as secretory and cellular compartmentation systems.

Identification of optimal crops for delivery will be another key to success. There will, of course, be instances where it will be preferable to bring wild species into domestication, but in general we will be looking to change something or add something to an already domesticated crop species to take advantage of efficient agronomic production systems. For instance, as described in other chapters, the rubber tree as a plantation crop has the potential to deliver recombinant proteins within an established harvesting and production chain. The provision of enabling technology for genetic manipulation of appropriate crop species will also be key to success. Provision of such specialized technology is not always the most glamorous scientific undertaking and may well require a certain amount of strategic vision to elicit supporting funds.

The identification and isolation of the relevant genes for the remodelling of pathways is a diminishing hurdle thanks to the pace of DNA sequence analysis of small plant genomes and the proliferation of EST (expressed sequence tag) libraries. Bioinformatic tools are well advanced in this arena for rapid access to tractable genetic information. Some would argue that small genomes are unlikely to provide access to the diversity of genes necessary for the remodelling of complex pathways. Against this we can rationalize diversity as the evolutionary product of a combinatorial approach in which a limited number of core enzymic functionalities are sequentially permutated. Genetic diversity in this context reflects conservatism manifest in small changes in the specificity of the core functionalities, leading to a complex array of gene families, the commonalities of which will be well represented even in small genomes. Information from small genomes will thus provide access to gene probes that can then be used to isolate the specific gene function required from less-well-characterized species. This argument applies equally well to both structural gene and regulatory gene families (e.g. transcription factors).

The identification of valuable functionalities, bridging the gap between commercial needs and opportunities and accessible biochemical diversity, remains a key factor. High throughput screening tools coupled to robotics is a compelling concept that is already being seriously embraced by the pharmaceutical industry in the quest for 'leads' from wild species. It is expected that other sectors will not be far behind in the development of this approach.

Conclusion

Having commented upon a wide range of influences or key factors for success both for the field in general and also for individual initiatives, I should emphasize that ultimately success will be dependent upon establishing the right balance between the sociopolitical aspects and technical opportunities. My own view is that success will grow best in microclimates or in partnerships that are buffered from the economics of the global market.

References

1. Bu'lock, J. and Kristiansen, B. (1987) Basic Biotechnology, Academic Press, New York

Antibody production in plants

C.P.E. van der Logt, C.S. Sidebottom and P.J. Davis*
Unilever Research, Colworth Laboratory, Colworth House, Sharnbrook, Bedford
MK44 1LQ, U.K.

Introduction

There is much to be gained from a new technology that allows the production of low-cost antibodies (or fragments of antibodies) on an industrial scale. Many different applications can be envisaged, some nearer to practical realization than others, all depending on the particular properties of specificity and affinity. But there is more to antibodies than is apparent at first sight, for they are amazing molecules, offering perfection in molecular recognition and incorporating a wide range of accessory functions through which to interface with the rest of the immune system. They are produced with high efficiency by plasma cells that develop as specialized protein-producing cells from precursor B lymphocytes. It is estimated that these cells produce 2000 antibody molecules per second at the peak of their productivity. The immune system builds antibodies that bind to a vast range of antigens with high affinity and specificity, and which trigger a variety of effector mechanisms. The challenge for the plant scientist is to develop an industrial-scale technology for the production of antibodies, allowing the wider exploitation of these remarkable molecules.

Antibodies

Antibody technology

Some of the special properties of antibodies have already been adapted and exploited in the laboratory and in medicine to achieve things that could not otherwise be done, providing unique analytical specificity and sensitivity for diagnostic immunoassays and, potentially, great therapeutic effect in certain diseases [1]. Consequently, antibodies are used in research, medicine, public health and, to a limited extent, some aspects of industry. Altogether, the existing practical uses of antibodies touch the lives of virtually the entire population. To underline the point, there are now home diagnostic tests for use by the general public that harness the power of antibodies in commercial consumer products, manufactured on an industrial scale and providing a medical benefit (e.g. personal fertility management).

The potential of antibodies has been enhanced by a succession of new technologies, the first of which was hybridoma technology, allowing the isolation

*To whom correspondence should be addressed.

of cell lines secreting antibodies of a single specificity [2]. This, together with increasingly powerful molecular biology and protein engineering techniques, led to the creation of antibodies and antibody-derived molecules designed to suit possible applications not only in the medical [1,3], but also in the industrial world. Antibody genes can now be expressed in various host organisms, either as whole immunoglobulins (Ig) or as active fragments tailored through genetic manipulation for particular applications [3]. The technology is now poised for more widespread application, but only if the costs and scale of production can be improved by orders of magnitude. A technology for the production of these amazing molecules in plants may provide those improvements, so unlocking the power of antibodies for new industrial uses.

Antibody structure

Plasma cells have certainly mastered the art of antibody production, being able to transcribe the genes, translate the mRNA, fold the proteins and assemble the entire structure with unparalleled efficiency – antibodies are their business! The task for the plant molecular biologist is now is to select and adapt plant hosts to achieve the same level of efficient, high fidelity production of antibodies or their fragments, but without the benefit of specialized plasma cells. The starting point is to understand the special molecular features and, hence, cellular demands of antibody proteins.

Several general features mark out antibodies as a distinct group of proteins [4]. Firstly, it must be recognized that they are complex molecules built in discreet modules called Ig domains or folds. Each domain consists of about 110 amino acids and, typically, the folding pattern is stabilized by an intrachain disulphide bond. Most of the sequence adopts a β-strand configuration, arranged into antiparallel strands, each of 5–10 amino acids, to create a sandwich of two β-sheets, often called a β-barrel [5]. All Ig domains have a hydrophobic interior, formed by in-pointing hydrophobic amino acids that alternate with out-pointing hydrophilic amino acids in the β-strands. The essential elements of this general architecture have been incorporated into an algorithm that can be used to objectively assign protein domains into (or out of) the Ig superfamily, through the computer program ALIGN [4], available as part of the NBRF database package (Protein Identification Resource – Protein Sequence Database, National Biomedical Research Foundation, Washington, D.C., U.S.A.).

Functional antibodies are assembled from a series of Ig domains to give a range of different models or classes, IgG, IgM, IgA, IgE and IgD in humans, which share the same general architecture and building blocks, but differ in their detail. IgG, the general 'workhorse' type of antibody, is the most abundant, and each intact IgG antibody molecule consists of two pairs of polypeptide chains, unhelpfully called 'light chains' and 'heavy chains'. The essential point is that the two light chains are identical and consist of two domains, while the two heavy chains are also identical and consist of four domains, as shown in Figure 1. Domains are designated as either constant (C) or variable (V), reflecting the extent of variation between domains present in different antibodies

The V-domains always occupy the N-terminal ends of the light and heavy chains and, as their name implies, they vary in their amino acid sequence

Figure 1

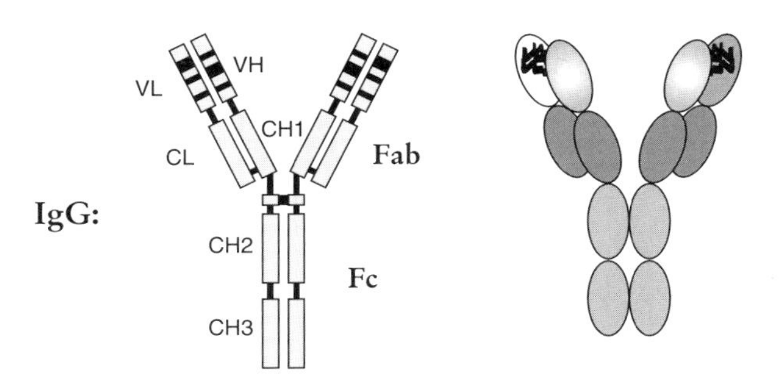

A diagrammatic representation of antibody structure

Two different forms of symbolic representation are depicted to show the modular construction of the intact molecules. The depiction on the left is a linear view in which each domain appears as a shaded rectangle, and is equivalent to an identifiable gene. In an immune response, the B-lymphocytes splice genes together to form an assembled, intact immunoglobulin gene. The antibody symbol on the right is a frequently used depiction which emphasizes (in a highly schematic form) the globular nature of each domain, each with an intrachain disulphide bond.

from one antibody to another. However, the variation is not wholly random and tends to occur within particular regions of the sequence. Within each V-domain, three hypervariable regions (complementarity determining regions or CDRs) are responsible for the antigen-binding specificity and affinity of the antibody. These CDR regions are flanked in the sequence by four relatively invariant framework regions (FR) regions. The FR regions form the antiparallel β-sheet structure that constitutes the structural scaffold of the V-domain. The VL and VH variable domains associate mainly through hydrophobic interactions between their β-sheets. This forms a barrel-like structure with all six CDRs exposed as loops. The VH–VL pair module that is responsible for the binding of antigen is referred to as the Fv of the antibody [4,5].

Three heavy-chain constant domains are present in IgG (CH1, CH2 and CH3). The CH1 and CH2 regions are connected via a flexible hinge region that is covalently linked to the hinge region of its identical partner heavy chain by disulphide bonds. Light chains have only one constant domain, which packs against the CH1 domain. The constant regions of the immunoglobulin are responsible for a variety of effector functions, such as complement fixation and opsonization for cells of the reticuloendothelial system.

It has recently been discovered that camels and their relatives posseses classes of antibodies that lack the light-chain counterpart of normal immunoglobulins, as shown schematically in Figure 2 [6]. These IgGs also lack CH1, which in one camel IgG class is structurally replaced by an extended hinge. Comparison of the sequence and structure of classical murine or human antibodies with these camel antibodies has shown that specific amino acid substitutions stabilize the structure and increase the solubility of the unpaired VH domain [7].

Antibodies, therefore, exist in various forms or sizes, and it is these complex, modular proteins, or active fragments of them, that are now to be

Figure 2

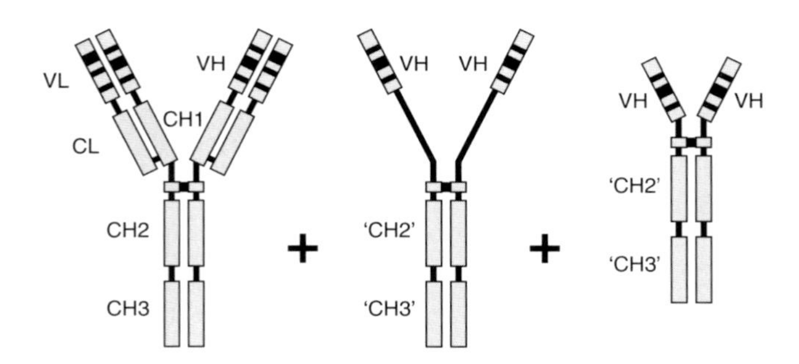

Diagrammatic representation of camel antibodies to illustrate their basic molecular architecture

Camels and their relatives have unusual forms of antibodies that may be useful in industrial applications. Normal IgG is always present, as well as the 'heavy-chain-only' versions. Each domain appears as a shaded rectangle, and is equivalent to an identifiable gene.

produced in engineered crop plants on an industrial scale, if the technology allows. The question is whether plant cells, which have not evolved to deal with the peculiarities of Ig proteins (in contrast to plasma cells), can ever be really efficient antibody producers, and to what extent genetic engineers can tailor the genes and select the host species to maximize productivity.

Antibody engineering

For many purposes most of the antibody molecule may not be necessary, or even desirable. For example, it has been found that repeated doses of rodent antibodies in human subjects elicit an anti-immunoglobulin response, referred to as human anti-mouse antibody or HAMA [8]. In addition, the relatively large molecular size of immunoglobulins can impair their tissue penetration while smaller, active fragments became the reagents of choice for some clinical applications that benefit from rapid clearance, since their circulating half-lives *in vivo* are much shorter than those of intact antibodies [9]. Antibody fragments, as depicted in Figure 3, can be prepared by enzymic cleavage of whole antibodies or via genetic engineering. Papain cleaves IgG antibody molecules in the hinge region and yields three fragments from each intact molecule: one fragment composed of the CH2 and CH3 domains (Fc fragment) and two identical fragments composed of the entire light chain combined with the VH–CH1 fragment (Fab fragment). After pepsin cleavage the two Fabs remain connected via the disulphide bonds in the hinge, thus forming the $F(ab')_2$ fragment.

The same domain structure that allowed the enzymic cleavage of whole antibodies also facilitates the isolation of the genes encoding the whole antibody or only the antigen binding (Fv) region [10,11]. Antibody genes have traditionally been isolated from hybridomas. In recent years the development of new strategies has allowed the construction of engineered antibody fragments directly from repertoires of antibody V-genes, without making hybridomas or even without the need for immunization [12,13]. The availability of the genes coding for the antibody and the recombinant expression of these genes in heterologous systems

Figure 3

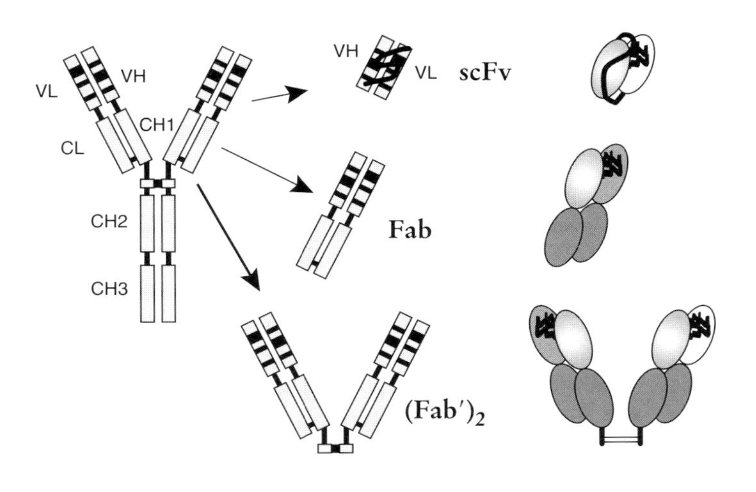

Diagrammatic representation of antibody fragments derived from IgG

Fab and F(ab')$_2$ can be derived by enzymic cleavage of whole IgG, but it is very difficult to release intact Fv in the same way. Single-chain (sc) Fv can only be produced by genetic engineering. The modular nature of the assembled immunoglobulin gene, in which each shaded rectangle represents an excisable domain gene, allows straightforward cloning of the various fragments by standard molecular biological techniques. It would be advantageous to produce in plants whole antibodies and/or each of the fragments depicted here.

also allows the tailoring of the antibody fragment to confer additional properties [14]. For example, peptide tails can be added [15], they can be fused with enzymes [16], two Fvs can be combined to construct bi-specific molecules [17–19] and targeting signals (sequences) can be added to direct expression of the antibody to a specific compartment of the host cell [20,21]. Expression of whole antibodies, Fabs and F(ab')$_2$s in heterologous systems is hindered by the fact that functional whole antibodies are only obtained by the correct assembly via covalent and non-covalent interactions of the heavy chains and the light chains [21]. This association takes place in the endoplasmic reticulum where chaperone proteins like BiP and the enzyme protein disulphide isomerase facilitate this assembly [22].

The construction of smaller antibody fragments does not require such stringent assembly conditions, thus also allowing expression in other cell compartments, which could offer many advantages. An interesting and valuable antibody fragment in this respect is the small single-chain antibody in which the variable domains of light and heavy chain are connected via a flexible linker chain, as depicted in Figure 3 [23]. Several problems inherent to the post-translational processing of whole antibodies can be circumvented by expression of single-chain (sc) Fvs. The scFv has therefore been the molecule of choice in many studies on the application and expression of antibody fragments. Paradoxically, however, it is often reported that in practice the best yields from transformed plants are with whole antibodies rather than small fragments [24].

Applications of antibodies

Antigen recognition and binding by antibodies and antibody fragments is exploited in numerous biological and medical applications [1,3,8,11]. Most of the widespread and well-established applications are '*ex situ*' applications in which isolated antibodies are used as tools in diagnosis, therapy and purification. Production cost is a major disadvantage of monoclonal antibodies, which has limited their wider use outside the medical area. With the advances in genetic engineering it is now possible to isolate the antibody genes and produce antibodies and antibody fragments at lower cost on a large scale in alternative hosts, thus opening the door for large-scale industrial applications.

The second area of antibody exploitation, which is much less well-developed, is the so-called '*in situ*' use, where a recombinant antibody acts as an effector agent in the cell that synthesized it. For example, there is much interest in using this approach to introduce pathogen protection into crop plants or to manipulate agronomic properties. Another '*in situ*' application of antibody fragments is a procedure also termed 'antisense protein strategy', in which antibodies are used to modify the *in vivo* activity of the antigen. The creation of dominant, loss-of-function mutant phenotypes by the expression of antibody genes in heterologous cells was first shown for yeast, using an antibody that bound and inhibited the activity of a cytoplasmic enzyme [25]. Subsequently, it was shown that accumulation of an anti-phytochrome scFv in transgenic tobacco plants was accompanied by aberrant phytochrome physiology in the transformants [26]. The growing list of examples of antibodies functioning as '*in situ*' agents clearly demonstrates the potential utility of antibodies for the modulation of intracellular antigens.

Hosts for antibody expression

Myeloma cells are the most widely used cells for production of whole recombinant antibodies. Very high expression levels can be obtained, especially after gene amplification selection. Yields of 500 $mg \cdot l^{-1}$ of recombinant antibody have been obtained in fed-batch air lift fermenters [27]. Alternative eukaryotic cell lines, such as Chinese hamster ovary (CHO) [28] and monkey kidney cells (COS) [29], have also been shown to produce in excess of 200 mg of recombinant antibody per litre of culture supernatant.

In recent years baculoviruses have joined the list of expression systems for the high-level production of recombinant antibodies and antibody fragments [30]. For many years the expression of antibodies in *Escherichia coli* has been a favourable alternative to animal cells for a number of reasons: vectors are available, growth is fast, the medium is cheap and there exists a wealth of expertise in the scale-up from laboratory-scale fermentations to production scale. Early attempts to produce whole antibodies in *E. coli* were disappointing, as the resulting polypeptides were compartmentalized in an insoluble form in inclusion bodies [31]. These problems were overcome by fusing the antibody genes to a bacterial signal sequence that directed the expressed antibody fragment to the periplasmic space. Correctly folded and fully active Fab, Fv and scFv antibody fragments have been produced in *E. coli* in this manner [32,33]. This development led to the expression of an ever expanding family of antibody fragments.

Whole antibodies, Fab and scFv fragments have been expressed and secreted by baker's yeast *Saccharomyces cerevisiae.* However, the yields of functional, correctly folded antibody or antibody fragments were only micrograms per litre [34]. Recently, the methylotrophic yeast *Pichia pastoris* has attracted attention [35], as it has strong, stringently controlled promoters, high levels of expression (levels of >100 $mg \cdot l^{-1}$ have been reported for scFv), easy downstream processing and a glycosylation pattern closer to the mammalian type. These features, combined with extensive fermentation experience, make this yeast exceptionally attractive for small- and medium-scale production of antibodies and antibody fragments. Certain species of filamentous fungi (e.g. *Aspergillus, Trichoderma*) have also been shown to be able to secrete large quantities (up to 25 $g \cdot l^{-1}$) of recombinant protein. First reports on the production of antibody fragments are promising (150 $mg \cdot l^{-1}$ of Fab produced by *Trichoderma reesei*) [36] but, as yet, this group of eukaryotic hosts are much less widely exploited for the expression of antibody fragments than yeast and bacterial expression systems.

Expression of antibodies and antibody fragments in plants

Since the first demonstration that plant cells are able to synthesize and secrete antibodies [37], plants have been considered as very attractive alternative hosts for immunoglobulin expression [38,39]. Part of the difficulty with every recombinant host lies in the transfer and expression of two genes (heavy and light chains) and the subsequent assembly of the synthesized polypeptides. Significant advances have been made in the techniques for plant transformation, such that the introduction of foreign genes into a number of plant species, including many economically important crop plants, is now routine. These advances, together with the fact that plants are generally efficient in processing complex proteins, have made plants very promising candidates for the expression of heterologous recombinant proteins, including antibodies and antibody fragments [40]. Glycosylation takes place in a manner that helps stability and function, but is different to that which takes place in mammalian cells, which could pose a problem for *in vivo* use.

Bulk production of antibody fragments

Plants have many potential advantages over production systems based on microbial fermentation or animal cell culture. Relatively efficient and straightforward gene transfer techniques are available for a wide range of species [40,41]. Antibodies and antibody fragments are normally expressed in plants with the correct three-dimensional folding and are usually active and very stable. The expression of recombinant antibodies can be directed to different parts of the plant for efficient accumulation and storage [42–44]. Antibodies can be secreted via the endoplasmic reticulum and Golgi apparatus to the intercellular apoplastic space, where they accumulate more efficiently, with reduced turnover and proteolysis [45,46].

Microbial fermentation and animal cell culture require huge capital investments and, in the case of animal cell culture, expensive growth media. In plants, the upstream production costs are much lower than for other systems, and they are probably the most cost-effective way of producing biomass on a large scale. Since antibodies may be expressed in plants that are categorized as GRAS (generally regarded as safe) organisms, no major regulatory hurdles associated with the host are expected. Also, the infrastructure for harvesting, transport and processing of plant biomass at very large scale is in place or can be easily adapted. Table 1 lists some estimates of the yield of antibody fragment per hectare, the amount of raw material that must be processed to produce one kg of antibody or antibody fragment and the projected production cost per kg of antibody or antibody fragment. The various crop plants listed are those that are being used or considered for use as hosts for recombinant antibody production. It should be noted that all the production costs given are rough estimates and only cover production of the recombinant antibody in the plant. The further costs arising from the required downstream processing are not included. The economics of extraction and downstream processing from plant biomass have not yet been investigated in any detail, except for the industrial processing of alfalfa [47].

If purification is necessary, it is essential that the downstream processing of the recombinant protein is economically acceptable, for the benefits of plant-based production should not be offset by increased downstream processing costs. Downstream processing from plant biomass is generally assumed to be difficult and expensive, because of the low ratio of recombinant protein to total biomass, so there are real incentives for increasing expression efficiency. Since there are no examples of single proteins extracted from crop plants on a large commercial scale, there are no benchmarks for estimating the likely costs. However, the antigen binding activity of antibodies and antibody fragments is a fundamental property that can allow the purification and concentration by a one-step affinity chromatography process, so reducing this assumed disadvantage considerably. This approach of using tailored epitopes (antigen fragments) is currently the subject of research in several groups.

Table 1

Crop	Yield of Fv ($kg \cdot ha^{-1} \cdot yr^{-1}$)	Production cost of Fv (£$\cdot kg^{-1}$)	Biomass ($tonnes \cdot kg^{-1} \cdot Fv^{-1}$)
Seed: rapeseed*	2.6	75	0.5
Seed: linseed†	1.4	100	0.5
Oleosins (Hirudin)	5.0	70	0.5
Leaf: tobacco	3.0	700	2–5
Tuber: potato¶	5.0	1200	10

*,†,¶ *Figures are based on the assumption of the heterologous protein being expressed at 1% of total seed protein or 1% of total tuber protein.*

Estimated yields and production costs of antibody fragments expressed in different plants

The production costs only consider the production of the antibody in the plant. Further costs will depend on the extent of the downstream processing required.

Antibodies in protection against pathogens

Antibodies directed against essential viral proteins can inactivate (or neutralize) viruses by agglutination or blocking of virulence epitopes. When these antibodies are expressed *in situ* they can operate as an effector mechanism that reduces pathogenicity, thus providing an alternative approach through which to engineer viral resistance [48]. For example, Voss and co-workers. [49] reported an important proof of principle in which virus resistance was imparted by the expression of a secreted whole antibody in tobacco. The antibody used in this study was 'neotope'-specific [binding to tobacco mosaic virus (TMV) via an epitope found only on assembled virus] and had already been shown to almost completely inhibit the infectivity of TMV when premixed prior to inoculation. A one-step transformation strategy yielded transgenic tobacco plants containing full-size cDNAs (including the native signal sequences) of both heavy and light chains of the antibody. The resulting transgenic tobacco plants expressed correctly assembled functional antibodies that were exported to the intercellular space. Upon infection with TMV the transgenic antibody-producing plants showed a significant reduction in the number of necrotic lesions that correlated with the amount of antibody produced. Although this study clearly demonstrated the importance of high expression levels, future work will also focus on targeting the antibodies in order to neutralize pathogens or alter pathogen function by immunomodulation. Efficient ways of targeting these molecules to different subcellular compartments have the potential to significantly increase the efficacy of protection, because many steps in pathogenesis are highly compartmentalized. Interfering with the intracellular mechanism of pathogenesis has, therefore, the potential to overcome problems associated with coat-protein-mediated protection.

Antiviral antibodies retained within the cytoplasm, rather than being secreted, were shown by Tavladoraki and co-workers to protect against virus infection [50]. They were able to show that a cytoplasmically targeted scFv against the artichoke mottled crinkle virus (AMCV) was constitutively expressed in tobacco plants and accumulated at low levels in the cytoplasm. The transgenic plants were measurably less susceptible to infection by the target virus and the appearance of viral symptoms was significantly delayed in comparison with untransformed plants. How the intracellular 'plantibody' confered protection against the virus is unknown, but this study has provided another clear proof of principle.

Modulation of plant antigens by *in situ* antibodies

Recombinant antibodies have been introduced into plants to alter or interfere with the function of plant antigens and to block phytohormones and metabolites. Phytohormones are a good example of target molecules that are not able to be effectively controlled through the antisense approach. Alternatives to antisense intervention are needed if, for example, the target molecule is encoded by a gene that has not been cloned, or if the particular target is a member of a large gene family (in which case the antisense gene could silence the whole set). Antibodies, especially single-chain antibody fragments expressed within an appropriate cellular compartment, might be the ideal alternative, for their fine selectivity can allow a target molecule to be inactivated or modulated with absolute specificity. A first proof of principle for this type of plant antigen modulation by *in situ* antibodies was provided by Owen and co-workers [26]. A scFv, directed against a conserved epitope of the regulatory photoreceptor phytochrome A, was expressed in the cytoplasm of tobacco cells. Those plants that expressed the scFv gene accumulated low but extractable levels of active antibody fragment. Phytochrome-dependent germination and photocontrol of hypocotyl elongation were both aberrant in the transgenic plants, indicating that the anti-(phytochrome A) scFv was not only present but also was able to interfere with phytochrome A function *in vivo.* A separate study confirmed the general feasibility of the concept by demonstrating that scFv antibodies that bind the phytohormone abscisic acid could also interfere with the target function when expressed in tobacco cells [51].

More speculative is the use of antibodies as intracellular anchors, achieved by linking the antibody genes to structural elements in the cell or to suitable trafficking signals. In such a format, antibodies could either capture the antigen or divert it from its normal location, possibly by targeting it to degradative compartments – effectively suicide antibodies.

Overall, this kind of intracellular immunization parallels antisense-RNA strategies, but has the added advantage that it can be applied to interfere with non-protein antigens and post-translational modification of proteins. It thus provides a versatile and promising experimental strategy for research, allowing the modification of raw materials and crop development through the binding of virtually any antigen within a plant, whether of plant or non-plant (e.g. viral) origin.

Species of plants for antibody production

A wide range of plant species, including tobacco, rapeseed, linseed, potato and, more recently, also the rubber tree (see Yeang, H.Y. and co-workers in Chapter 6 of this volume), are considered to be potential hosts for industrial antibody production. Whereas it is important to select the right host species for the application, it is equally important to select the best part of the plant for accumulation and storage of the recombinant antibody. The main tissues used (or proposed for use) for the accumulation of recombinant antibody are tubers, leaves and seeds.

Tubers

Potato tubers are an obvious candidate for the large-scale production of any recombinant protein due to their high yield (30 tonne·ha^{-1}) although their protein content is low. In addition, the infrastructure for growing and harvesting large amounts of potato tubers is well-established and it would be possible to utilize commercial equipment for processing the tubers and extracting the recombinant protein. Even so, there are few examples of tubers being used as expression systems for heterologous proteins and, in those examples that do exist, the levels of expression are quite low. For example, human serum albumin was expressed at 0.02–0.04% of total protein. Active antibody fragments have also been expressed in transformed potatoes in our laboratory, providing the first example of antibodies expressed in tubers. As with human serum albumin, the antibody fragment expression levels were still too low to make production in potato tubers economically acceptable. It will be necessary to increase expression levels to at least 1% of total protein (see Table 1). Despite these problems, potatoes are one of the major transgenic crops being grown in Europe [52,53], but not yet as commercial factories for heterologous proteins.

The fact that stored tubers remain metabolically active is a potential problem with the use of potatoes. Any industrial-scale process would depend on antibody fragments remaining stable throughout storage and processing of the tuber. Preliminary results using transgenic potatoes expressing the anti-*Streptococcus sanguis* antibody 4715 and potato extracts spiked with model antibody fragments look very promising. Incubation of antibody fragments in potato extracts for up to 16 h did not affect the binding activity of two out of three antibody fragments tested. The third antibody fragment was shown to be stable for up to 4 h in potato tuber extract but it quickly lost all binding activity upon longer incubation. The commercial development of antibody production in potato tubers will require selection of stable antibody fragments, as well as optimization of the extraction process.

Leaves

A number of groups are working on the expression of proteins, including antibodies and antibody fragments, in leaf tissue. The expression system most often used is tobacco, as this species is readily transformed and regularly used as a model system. For example, in the pioneering work of Hiatt and colleagues, tobacco was used to express an anti-*Streptococcus mutans* whole IgA antibody in the leaves [24]. There is a serious intention to scale-up production to commercially viable quantities, with tobacco and alfalfa leaf as the target crops. Yields of 3 kg of antibody per hectare at a cost of US1\cdot$g^{-1} are anticipated. Methods for the extraction of proteins from leaf tissue on a large scale already have been designed by Pirie and co-workers in the 1960s [54]. They developed a fully integrated system for high protein recovery from leaves, based upon pulping machines and sugar cane roll presses for the separation of the leaf sap containing protein from the fibrous pulp.

Seeds

Seeds are the parts of the plant most frequently targeted for heterologous protein expression. They have the advantage of being high-protein (e.g. oilseed rape has a protein content of 20% of the total dry weight), low-water systems in which expressed proteins are stable for up to 18 months [55]. In addition, seeds are one of the best characterized plant systems in terms of gene expression, while the processes of milling and protein extraction are already well-developed for a number of crops. However, there are also a number of disadvantages linked to the use of seed crops, such as the risk of cross-fertilization between different varieties. In France and Canada, where field trials have been carried out, transformed rape plants had to be physically contained. It is likely that, eventually, zonal planting of the transgenic varieties will be necessary. Dedicated processing facilities with specialized equipment will have to be developed, because the proteins currently expressed can be supplied by a relatively small acreage of plants on a scale that is not sufficiently large for traditional processing factories.

More recently the use of oleosins, one of the most abundant proteins in the seeds of oil-producing plants, is being explored for the expression of antibody fragments and other recombinant proteins (see Moloney, M.M. in Chapter 5). These proteins are uniquely associated with the oil-storage bodies, where they form a proteinaceous mono-molecular layer around the oil droplets [56]. Oleosin fusion proteins have been shown to accumulate up to 10% of the total cell protein [57]. This high level of expression of the recombinant protein, combined with the advantage of easy extraction, make the system a strong candidate for the bulk production of antibodies at low cost.

A worked example: production and stability of an antibody fragment with significant commercial potential

Antibodies used either directly, or as a targeting system for killing complexes, are being evaluated for their inclusion in dental products. Their use would provide a specific and more natural method of plaque control than the active chemicals currently used. The work of Ma and co-workers with the anti-*Streptococcus mutans* antibody Guy's 13 [24] is a good example of the potential use of antibodies in the fight against plaque. This work showed that topical application of the monoclonal antibody (mAb) inhibited caries development in both sub-human primates and humans, apparently by preventing the colonization of the teeth by *S. mutans*. Another oral bacterium, *S. sanguis*, is the predominant organism in early dental plaque (12–24 h). Although it is not thought to be directly pathogenic, *S. sanguis* creates a surface to which other cariogenic organisms can attach. It thus forms the foundation for plaque development on teeth and was therefore selected as a target organism for antibody-targeted delivery of cytotoxic agents. The variable region genes of an anti-*S. sanguis* mAb (designated 4715) were isolated and expressed in microbial systems. Both the scFv fragment and the parental whole antibody were shown to be very effective in targeted delivery of antibacterial enzymes and rapid killing of targeted cells (Figure 4).

Having established the anti-bacterial potential of the mAb 4715, we formatted the antibody fragment gene to allow expression in plants. The 35 S CaMV (cauliflower mosaic virus) promoter and NOS (nopaline synthase) terminator regulatory sequences and the selectable marker NPTII (neomycin phosphotransferase II) from the pGPTV binary vector, together with the coding region of the tobacco pathogenesis-related protein PR1a signal sequence [58], were used to direct expression of the scFv fragment to the apoplastic space (Figure 5. The inclusion of a peptide tail ('hydrophil-2') at the C-terminus of the light chain allowed the detection of antibody fragment and antigen binding activity produced in crude plant extracts. The scFv expression cassette was introduced into tobacco and potato by *Agrobacterium*-mediated transformation. Independent kanamycin-resistant transformed tobacco plants were screened for the presence of *S. sanguis* binding activity.

Twenty-nine out of 96 plants tested were shown to contain the antibody gene and to express detectable levels of *S. sanguis* binding activity. By comparing the staining intensity on Western blots and the reactivity in ELISA of the scFv

Figure 4

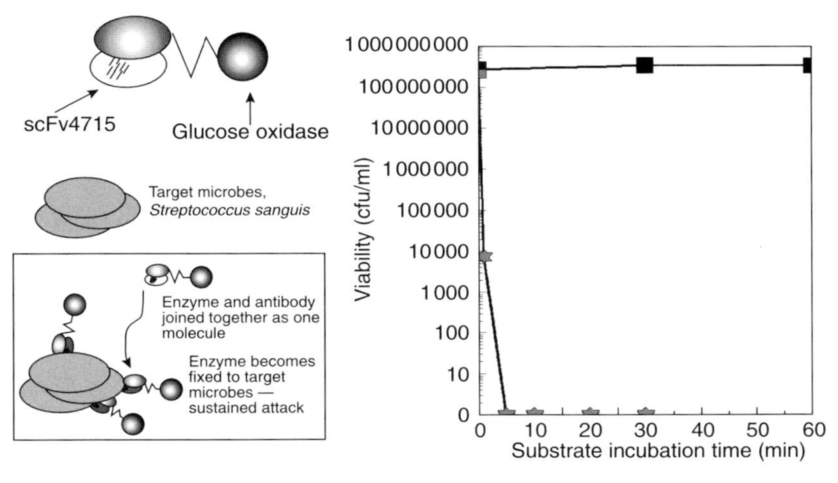

Microbial cell killing as an example of a potential large-scale use of engineered antibodies, relevant to almost the entire population

An antibody fragment specific for Streptococcus sanguis was chemically conjugated to the enzyme glucose oxidase. The conjugate was incubated with target S. sanguis cells and then unbound reagent was washed away. A set of control S. sanguis cells was treated in the same way with an equal amount of glucose oxidase not conjugated to an antibody fragment. The treated, washed cells were resuspended in a solution of glucose and sodium iodide with trace levels of peroxidase. The hydrogen peroxide produced by the glucose oxidase captured on the surface of the target microbes was sufficient to achieve a rapid kill (as shown on the graph) in the presence of iodide and peroxidase.

produced in the plant with known amounts of bacterially produced scFv, it was estimated that the maximum expression level reached was only 0.006% of total protein. To determine the expression of the antibody gene in the transgenic plants, total RNA was isolated from eight selected plants. Northern blot analysis clearly showed that all the selected plants expressed very high levels of single-chain antibody transcripts (Figure 6). In an attempt to increase the level of antibody production and accumulation, we self- and cross-pollinated some of the high-producer plants. Out of the 800 F1 plants analysed for antibody fragment expression, 711 plants were shown to be positive, expressing similar or up to 4-fold higher levels of antigen binding activity compared with the Fo plants. Expression levels of scFv 4715 in tubers of potatoes transformed with the same gene construct were also very disappointing, with expression levels not exceeding the 0.002% level observed for transgenic tobacco plants.

Future prospects for antibody production in plants

Although there have been impressive reports of the production of functional antibodies and antibody fragments in different cell compartments, experience from our own and other laboratories indicates that this is not always the case. For reasons yet unknown, some scFv fragments are expressed at levels up to 4.8% [51]

Figure 5

The plasmid used with *Agrobacterium*

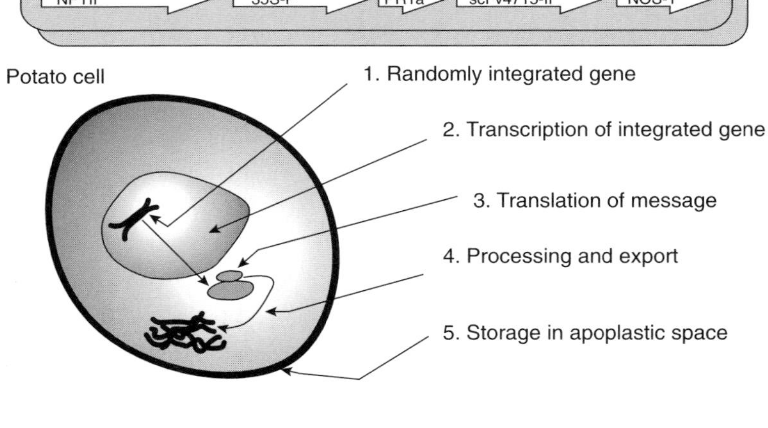

INTERVENTION POINT	1. Randomly integrated gene	2. Transcription of integrated gene	3. Translation of message	4. Processing and export	5. Storage in apoplastic space
STRATEGY FOR IMPROVEMENT	Targeted integration into already transcribed region	Specific promoter and enhancer sequences	Optimal codon usage and RNA stabilizing elements	Molecular chaperones, targeting sequences etc.	Alternative storage compartments

Opportunities for gaining improvements in the expression of heterologous genes in plant cells

The expression system used at present is shown diagrammatically to highlight those aspects that are considered to be less than optimal. If substantial improvements in yield are to be gained, then at least some of the modifications indicated in the Figure will be needed.

of total soluble protein while other antibody fragments do not accumulate to detectable levels, even if they are targeted to different cellular locations, including the endoplasmic reticulum, which has been shown to enhance the accumulation of some scFvs.

Ma, Hiatt and co-workers have observed in their work on the expression of anti-*S. mutans* IgA in tobacco plants [24], that one of the key factors influencing expression levels is the inherent stability of the active conformation. They initially expressed heavy and light chains in separate plants that were subsequently cross-pollinated. The progeny was shown to have levels of RNA very similar to those of the parents, but the levels of active whole antibody in the leaves increased by up to 1000-fold. They also achieved higher levels of accumulation with whole antibody rather than with Fab or Fv, suggesting that the complexity of the antibody assembly has a synergistic effect on the expression.

The key to the successful exploitation of plants for the production of antibodies and antibody fragments, or even for the exploitation of antibodies on an industrial scale in general, is to increase the expression of recombinant antibody to an economic level. Although several different strategies have been proposed to achieve this aim, no integrated approach has been documented. It is in

Figure 6

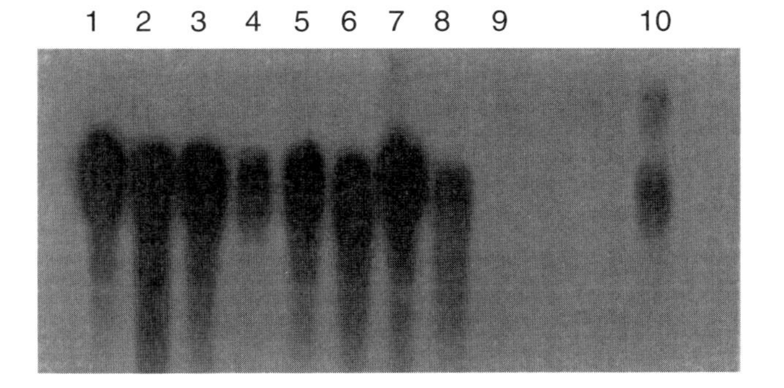

Northern blot analysis of transgenic potato plants carrying the genetic construct encoding scFv 4715

A group of 29 potato plants was shown to contain the scFv gene and to express detectable levels of S.sanguis binding activity. However, the total and active levels of scFv 4715 were estimated to be only 0.006% of total protein. The expression of the antibody gene in the transgenic plants was assessed by isolating total RNA from eight selected plants (lanes 1–8). Lane 9 is RNA from an equivalent untransformed plant, while lane 10 is RNA from the parent hybridoma line. The Northern blot analysis of the total RNA extracted from 8 selected plants shown in this Figure is clear evidence that all selected plants expressed very high levels of single-chain antibody transcripts, and confirms that the problem to be solved is further down in the protein synthesis system of the potato cell (see Figure 5).

this area that major advances need to be made, and it is most likely that further development of the gene constructs will enhance expression of any recombinant protein in transgenic plants. To maximally exploit crop plants it will be necessary to tailor the genetic constructs for use in particular plants, as depicted in Figure 5. This will require the use of appropriate promoter, enhancer and leader sequences, together with optimized codon usage and the removal of mRNA destabilizing sequences. Further improvements could come from integration-independent expression and targeting to specific cell compartments. Optimization of this type is not easy to achieve and, at this stage in the development of the technology, poses more questions than it answers. However, the requirements for achieving higher expression levels can set the agenda for future research and development of crop plants as antibody production systems.

References

1. Haber, E. (1992) Immunol. Rev. **130**, 189–212
2. Kohler, G. and Milstein, C. (1975) Nature (London) **256**, 495–497
3. Verhoeyen, M.E. and Windust, J. (1995) in Molecular Immunology (Hames, B.D. and Glover, D.M., eds.), pp. 283–325, IRL Press, Oxford
4. Williams, A.F. and Barclay, A.N. (1989) in Immunoglobulin Genes (Honjo,T., Alt, F.W. and Rabbitts, T.H., eds.), pp. 361–387, Academic Press, London
5. Amzel, M. and Poljak, R.J. (1979) Annu. Rev. Biochem. **48**, 961–997
6. Hamers-Casterman, C., Atarhouch, T., Muyldermans, S., Robinson, G., Hamers, C., Bajyana Songa, E., Bendahman, N. and Hamers, R. (1993) Nature (London) **363**, 446–448
7. Muyldermans, S., Atarhouch, T., Saldanha, J., Barbosa, J.A.R.G. and Hamers, R. (1994) Protein Eng. **7**, 1129–1135
8. Verhoeyen, M.E.V., Saunders, J.A., Broderik, E.L., Eida, S.J. and Badley, R.A. (1991) Dis. Markers **9**, 197–203

9. Davies, J. and Riechmann, L. (1996) Protein Eng. **9**, 531–537
10. King, D.J., Mountain A., Adair, J.R., Owens, R.J., Harvey, A., Weir, N., Proudfoot, K.A., Phipps, A., Lawson, A. and Rhind, S.K. (1992) Antibody Immunoconj. Radiopharm. **5**, 159–170
11. Huston, J.S., McCartney, J., Tai, M.-S., Mottola-Hartshorn, C., Jin, D., Warren, F., Keck, P. and Opperman, H. (1993) Int. Rev. Immunol. **10**, 195–217
12. Marks, J.D., Hoogenboom, H.R., Griffiths, A.D. and Winter, G. (1992) J. Biol. Chem. **267**, 16007–16010
13. Marks, J.D., Hoogenboom, H.R., Bonnert, T.P., McCafferty, J., Griffiths, A.D. and Winter, G. (1991) J. Mol. Biol. **222**, 581–589
14. Brinkmann, U., Buchner, J. and Pastan, I. (1992) Proc. Natl. Acad. Sci. U.S.A. 3075–3079
15. Ford, C.F., Suominen, I. and Glatz, C.E. (1991) Protein Exp. Purif. **2**, 95–99
16. Neuberger, M.S. Williams, G.T. and Fox, R.O. (1984) Nature (London) **312**, 604–607
17. Better, M., Chang, C.P., Robinson, R.R. and Horwitz, A.H. (1988) Science **240**, 1041–1043
18. Mallender, W. and Voss, E.W. (1994) J. Biol. Chem. **269**, 199–206
19. Holliger, P., Prospero, T. and Winter, G. (1993) Proc. Natl. Acad. Sci. U.S.A. 6444–6448
20. Skerra, A. and Pluckthun, A. (1988) Science **240**, 1038–1041
21. Fischer, G. and Schmid, F.X. (1990) Biochemistry **29**, 2205–2212
22. Lorimer, G.H. (1992) Curr. Opin. Struct. Biol. **2**, 26–34
23. Bird, R.E., Hardman, K.D., Jacobson, J.W., Johnson, S., Kaufman, B.M., Lee, S.-M., Lee, T., Pope, S.H., Riordan, G.S. and Whitlow, M. (1988) Science **242**, 423–426
24. Ma, J.K., Hiatt, A., Hein, M., Vine, N.D., Wang, F.F., Stabila, P., Van Dolleweed, C., Mostov, K. and Lehner, T. (1995) Science **268**, 716–719
25. Carlson, J.D. (1988) Mol. Cell. Biol. **8**, 2638–2646
26. Owen, M., Gandecha, A., Cockburn, B. and Whitelam, G. (1992) Biotechnology **10**, 790–794
27. Bebbington, C.R., Renner, G., Thomson, S., King, D., Abrams, D. and Yarranton, G.T. (1992) Biotechnology **10**, 169–181
28. Bebbington, C.R. (1991) Methods **2**, 136–142
29. De Sutter, K., Feys, V., Van De Voorde, A. and Friers, W. (1992) Gene **113**, 223–230
30. Hasemann, C.A. and Capra, J.D. (1990) Proc. Natl. Acad. Sci. U.S.A. **87**, 3942–3948
31. Cabilly, S., Riggs, A.D., Pande, H., Shively, J.E., Holmes, W.E., Rey, M., Perry, L.J., Wetzel, R. and Heynekker, H.L. (1984) Proc. Natl. Acad. Sci. U.S.A. **81**, 3273–3277
32. Better, M., Chang, C.P., Robinson, R.R. and Horwitz, A.H. (1988) Science **240**, 1041–1043
33. Skerra, A. and Pluckthun, A. (1988) Science **240**, 1038–1040
34. Horwitz, A.H., Chang, C.P., Better, M., Hellstrom, K.E. and Robinson, R.R. (1988) Proc. Natl. Acad. Sci. U.S.A. **85**, 8678–8682
35. Ridder, R. Schmitz, R., Legay, F. and Gram, H. (1995) Biotechnology **13**, 255–260
36. Nyyssonen, E., Penttilla, M., Harkki, A., Saloheimo, A., Knowles, J.K.C. and Keranen, S. (1993) Biotechnology **11**, 591–595
37. During, K. (1988) PhD Thesis, Universitat Koln, Germany
38. During, K., Hippe, S., Kreuzaler, F. and Schell, J. (1990) Plant Mol. Biol. **15**, 281–293
39. Hiatt, A., Cafferkey, R. and Bowdish, K. (1989) Nature (London) **342**, 76–87
40. Conrad, U. and Fiedler, U. (1994) Plant Mol. Biol. **26**, 1023–1030
41. Kareiva, P. (1993) Nature (London) **363**, 580–581
42. Fiedler, U. and Conrad, U. (1995) Biotechnology **13**, 1090–1093
43. van Engelen, F.A, Schouten, A., Molthoff, J.W., Roosien, J., Salinas, J., Dirkse, W.G., Schots, A., Bakker J., Gommers, F.J., Jingsma, M.A., Bosch, D. and Stiekema, W.J. (1994) Plant Mol. Biol. **26**, 1701–1710
44. De Wilde, C., De Neve, M., De Rycke, R., Bruyns, A., De Jaeger, G., Van Montagu, M., Depicker, A. and Engler, G. (1996) Plant Sci. **114**, 233–241
45. Bednarek, S.Y. and Raikhel, N.V. (1992) Plant Mol. Biol. **20**, 133–150
46. Firek, S., Draper, J., Owen. M.R.L., Gandecha, A., Cockburn, B. and Whitelam, G.C. (1993) Plant Mol. Biol. **23**, 861–870
47. Austin, S (1994) Ann. N. Y. Acad. Sci. **721**, 234–244
48. Schots, A., De Boer, J., Schouten, A., Roosien, J., Zilverntant, J.F., Pomp, H., Bouwman-Smits, L., Overmars, H., Gommers, F.J., Visser, B., Stiekema, W.J. and Bakker, J. (1992) Neth. J. Plant Pathol. **98**, 183–191
49. Voss, A., Niersbach, M., Hain, R., Hirsch, H.J., Liao, Y.C., Kreuzaler, F. and Fischer, R. (1995) Mol. Breeding **1**, 39–50
50. Tavladoraki, P., Benvenuto, E., Trinca, S., De Martinis, D., Cattaneo, A. and Galeffi, P. (1993) Nature (London) **366**, 469–472
51. Artsaenko, O., Peisker, M., Zur Nieden U., Fiedler, U., Weiler, E.W., Muntz, K. and Conrad, U. (1995) The Plant J. **8**, 745–750

52. Sijmons, P.C., Dekker, B.M.M., Schrammeijer, B., Verwoerd, T.C., Van den Elzen, P.J.M. and Hoekena, H. (1990) Biotechnology **8**, 217–221
53. Haq, T.A., Mason, H.S., Clements, J.D. and Arntzen, C.J. (1995) Science **268**, 714–716
54. Pirie, N.W. (1971) in Leaf Protein, its Agronomy, Extraction, Quality and Uses (Pirie, N.W., ed.), IBP Handbook 20, Blackwell Scientific Publications, Oxford
55. Pen, J., Verwoerd, T.C., Von Paridon, P.A., Beudeker, R.F., Van den Elzen, P.J.M., Geerse, K., Van der Klis, J.D., Versteggh, H.A.J., Van Ooyen, A.J.J. and Hoekema, A. (1993) Biotechnology **11**, 811–814
56. Murphy, D.J., Cummins, I. and Kang, A.S. (1989) Biochem. J. **258**, 285–293
57. Parmenter, D.L., Boothe, J.G., Van Roijen, G.J.H., Yeung, E.C. and Moloney, M.M., (1995) Plant Mol. Biol. **29**, 1167–1180
58. Pfitzener, U.M. and Goodman, H.M. (1987) Nucleic Acid Res. **15**, 4449–4465

Antigen expression on the surface of a plant virus for vaccine production

V.E. Spall*, C. Porta*, K.M. Taylor*, T. Lin†, J.E. Johnson† and G.P. Lomonossoff*‡
*Department of Virus Research, John Innes Centre, Colney Lane, Norwich NR4 7UH, U.K. and †Department of Molecular Biology, The Scripps Research Institute, 10666 North Torrey Pines Road, La Jolla, CA 92037, U.S.A.

Summary

Epitopes have been expressed on the surface of the plant virus, cowpea mosaic virus (CPMV). CPMV-based chimaeras grow extremely well in plants and, in the majority of cases, high yields of modified virus can be easily purified. Chimaeras expressing epitopes from either human rhinovirus 14 (HRV-14) or from human immunodeficiency virus type 1 (HIV-1) were found to possess the antigenic properties of the insert. Furthermore, the HIV-1 chimaera stimulated anti-HIV-1 neutralizing antibodies when injected into mice. The atomic resolution structure of the chimaera containing an epitope from HRV-14 has been solved and predictions have been made as to how the chimaera could be modified to maximize immunogenicity. The ability to solve the three-dimensional structures of chimaeric virus particles indicates that CPMV is a viable system for structure-based design of peptide presentation.

Introduction

In recent years antigenic peptides from pathogens have been investigated for their use as potential vaccines. Free peptides have been found to have immunogenic properties but their immunogenicity can generally be enhanced by presentation of the peptides on a carrier protein [1]. Such epitope presentation or expression systems increase the lifespan of a synthetic peptide in the cellular environment and present the peptide in a manner that is more likely to be biologically active. Initial studies involved chemical cross-linking of a polypeptide to a range of carrier proteins [2], but subsequently many systems have been developed that are based on genetically linking epitopes to a self-assembling macromolecule [3,4]. The primary requisites for such systems are that the peptide must be prominently exposed on the surface of the macromolecule and that the insertion of the peptide must not

‡To whom correspondence should be addressed.

interfere with the self-assembly process. Potential carrier molecules that have been investigated include the coat proteins of simple viruses of bacteria [5–7], animals [8–10] and plants [11–14]. Linear, continuous epitopes are commonly expressed in such systems and the choice of the insertion site for such sequences has often been aided by knowledge of the three-dimensional structure of the macromolecule. For example, knowledge of the structure of poliovirus assisted the choice of the epitope insertion site in the construction of poliovirus chimaeras [15].

The work presented in this chapter concerns the expression of immunologically important peptides on the surface of the plant virus CPMV. The use of a plant virus as an expression system obviates the need to propagate the recombinant protein in liquid culture under sterile conditions. In addition, as plant viruses are not infectious in animal cells, a 'dead' vaccine based on a plant virus should be extremely safe. CPMV has a number of potential advantages for this use: the virus grows extremely well in its host, cowpea (*Vigna unguiculata*), and high yields (1–2 g virus per kg of infected tissue) can be purified. The purification of virus particles is straightforward, taking less than a day, and the virus is thermostable. Thus a vaccine based on CPMV might not require the refrigeration needed to store most current vaccines. It may be possible, therefore, to produce low-cost, stable vaccines from cowpea plants. To underline the feasibility and practicality of the approach, results from both immunogenicity and structural studies of antigens presented in the CPMV expression system have been included and discussed in this chapter.

Description of the CPMV expression system

CPMV is the type member of the comovirus group of plant viruses. Its genome consists of two separately encapsidated positive-strand RNA molecules of 5889 (RNA 1) and 3481 (RNA 2) nucleotides [16,17]. The RNAs each contain a single open reading frame and are expressed through the synthesis and subsequent processing of precursor polyproteins (for a review, see [18]). Both RNAs are required to cause an infection in plants. RNA 1 encodes the proteins involved in the replication of both viral RNAs while RNA 2 encodes proteins involved in the cell-to-cell movement and encapsidation of the virus. The structure of CPMV has been solved to atomic resolution [19,20] and details of the comovirus structure have been described by Lomonossoff and Johnson [21]. CPMV capsids contain 60 copies each of a large (L) and a small (S) coat protein arranged with icosahedral symmetry. The two capsid proteins are folded into three antiparallel β-barrel structures, forming one icosahedral asymmetric unit of the virus. In total, 180 β-barrel domains comprise the quasi $T=3$ ($P=3$) structure of CPMV (Figure 1). Each L protein consists of two β-barrel domains that are positioned close to the icosahedral 3-fold axes. Five copies of the S protein, each consisting of a single β-barrel domain, are arranged around the 5-fold axes. The strands of the β-barrels are linked to each other with loops of varied sequence; in the related animal picornaviruses these loops commonly form the antigenic sites of the virus. An analysis of the three-dimensional structure of CPMV suggested that the loops between the β-strands would be suitable sites for the insertion of epitopes, since these sequences are not involved in contacts between protein subunits. One of the loops of the S protein, the βB-βC loop, was found to be highly exposed on the capsid and was initially chosen for the addition of epitopes (Figure 1). A comparison of the sequence and conformation of this loop in three different comoviruses showed a high degree of variability, suggesting that different amino acid sequences could be inserted at this site without abolishing normal virus function [3]. This chapter is concerned exclusively with the properties of CPMV-based chimaeras containing insertions into this loop of the S protein. Recently, however, we have demonstrated that it is also possible to insert heterologous sequences into βC′-βC′′ loop of the S protein (Figure 1) while retaining virus viability.

Figure 1

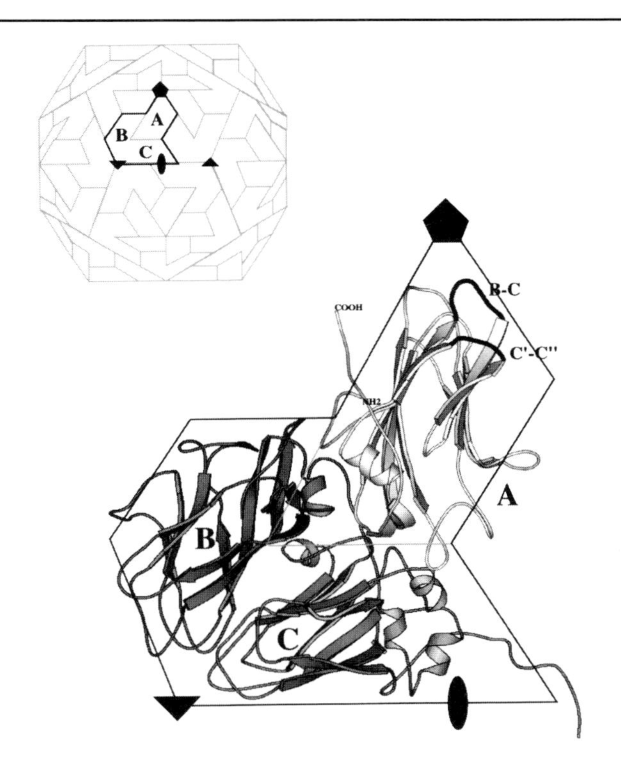

Quaternary organization of the large (L) and small (S) subunits of CPMV

The ribbon drawing represents the arrangement of the three β-barrels as they are thought to exist in the precleavage polyprotein and the heavy outline in the capsid model shows the location of this asymmetric unit in the particle. The S subunit consists of a single β-barrel domain (marked A) and is at the top of the asymmetric unit. The two loops that, to date, have been shown to tolerate the insertion of epitopes (B-C and C′-C′′) are highlighted in black. The L subunit is composed of two β-barrels (domains marked B and C) at the bottom of the drawing. The ribbon diagram has been altered slightly from the actual structure in that the angle between the L and S subunits has been increased so that the view is approximately into the barrel for all three domains. The icosahedral 5-fold, 3-fold and 2-fold axes are indicated by filled pentagons, triangles and ovals, respectively.

Full-length cDNA clones of both CPMV RNAs are available, allowing manipulation of the CPMV genome. The first CPMV chimaeras [11,12] were constructed using clones containing full-length cDNA copies of the two viral RNAs downstream of a T7 promoter. RNA transcribed *in vitro* from these clones was infectious when inoculated on to cowpea plants. Subsequently, plasmids that contain full-length cDNA copies of RNAs 1 and 2 downstream of the cauliflower mosaic virus 35 S promoter were constructed [22]. When linearized, these clones (termed pCP1 and pCP2) are directly infectious on cowpeas. These clones provide a cheaper, more efficient method for infecting plants than the use of *in vitro* transcripts, and are now used routinely for the construction of chimaeras.

Early investigations into the use of CPMV as an expression vector determined certain guidelines for the construction of viable, genetically stable, chimaeras [12]. Firstly, foreign sequences should be inserted as additions to the wild-type CPMV sequence and not used as replacements for native residues. Secondly, methods for introducing foreign sequences that result in sequence duplications either side of the insert are unsuitable, as loss of the insert occurs on passaging as a result of homologous recombination. Thirdly, the precise site of insertion is important in maximizing the growth of the chimaeras. Taking these guidelines into consideration, we have now refined a standard protocol for the introduction of foreign sequences into the βB-βC loop of the S protein. This involves inserting DNA fragments encoding the foreign sequence of interest between the unique *Nhe*I and *Aat*II sites of pCP2. Complementary oligonucleotides are synthesized that contain the sequence for the heterologous insert flanked by CPMV-specific residues and terminating in *Nhe*I- and *Aat*II-compatible ends. The oligonucleotides are phosphorylated, annealed and ligated into *Nhe*I/*Aat*II-digested pCP2. This one-step cloning procedure allows the direct insertion of the foreign sequence without loss of any CPMV-specific residues. By this method the heterologous sequence is placed immediately upstream of Pro-23 of the S protein, an amino acid that is highly exposed on the surface of the CPMV capsid. This method of construction is illustrated for the case of a chimaera containing an epitope from HRV-14 (Figure 2).

To propagate the chimaeras, modified pCP2 (pCP2 chimaera) and pCP1 are linearized and inoculated on to cowpea plants (Figure 3). Once an infection is established (usually by 10 days post-inoculation), infected leaves are taken and the virus is extracted by the standard CPMV purification protocol [23].

To date, more than 50 chimaeras have been produced in the manner described above. In the majority of cases the presence of the heterologous sequence does not affect the ability of the virus to grow in plants and, in most cases, the yields of modified virus are similar to those obtained from plants infected with wild-type CPMV. The infections can be sap-transmitted to healthy plants allowing the efficient propagation of chimaeric virus. An investigation into the genetic stability of a number of chimaeras has determined that inserts can be maintained intact through at least ten serial passages, provided the size of the insert does not greatly exceed 30 amino acids.

Figure 2

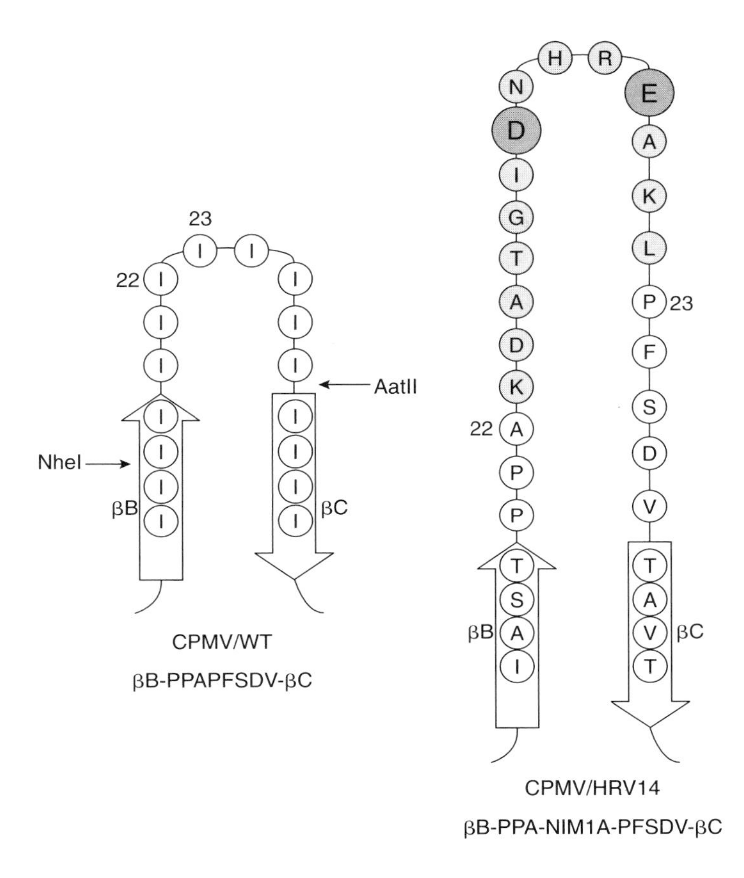

Structure of the βB-βC loops of the S protein in wild-type CPMV (left) and the chimaera HRV-II (right)

The NheI and AatII restriction sites used in the construction of the chimaeras are indicated. The residues corresponding to the inserted sequence from VP1 of HRV-14 are shaded, with the larger circles indicating the amino acids that define the NIm-1A site. Repinted from [3] by permission of the publisher, Academic Press Limited, London.

Antigenic and immunological properties of chimaeras

Among the first sequences to be expressed on the surface of CPMV were epitopes from HRV-14 and HIV-1. These chimaeras (termed HRV-II and HIV-III) contain residues 85–98 from VP1 of HRV-14 (the NIm-1A site [24]) and residues 731–752 from gp41 of HIV-1 [25], respectively. When the chimaeras were purified and the virions analysed by SDS-polyacrylamide gel electrophoresis, S proteins of the appropriately increased size were observed. In addition, a smaller protein (S′), which is not seen with wild-type virus, could be observed (Figure 4). N-terminal sequence analysis of the S′ protein from HIV-III showed that this protein arose through a proteolytic cleavage event between the last two residues of the insert [26]. A similar cleavage has also be found in the case of the S protein from HRV-II

Figure 3

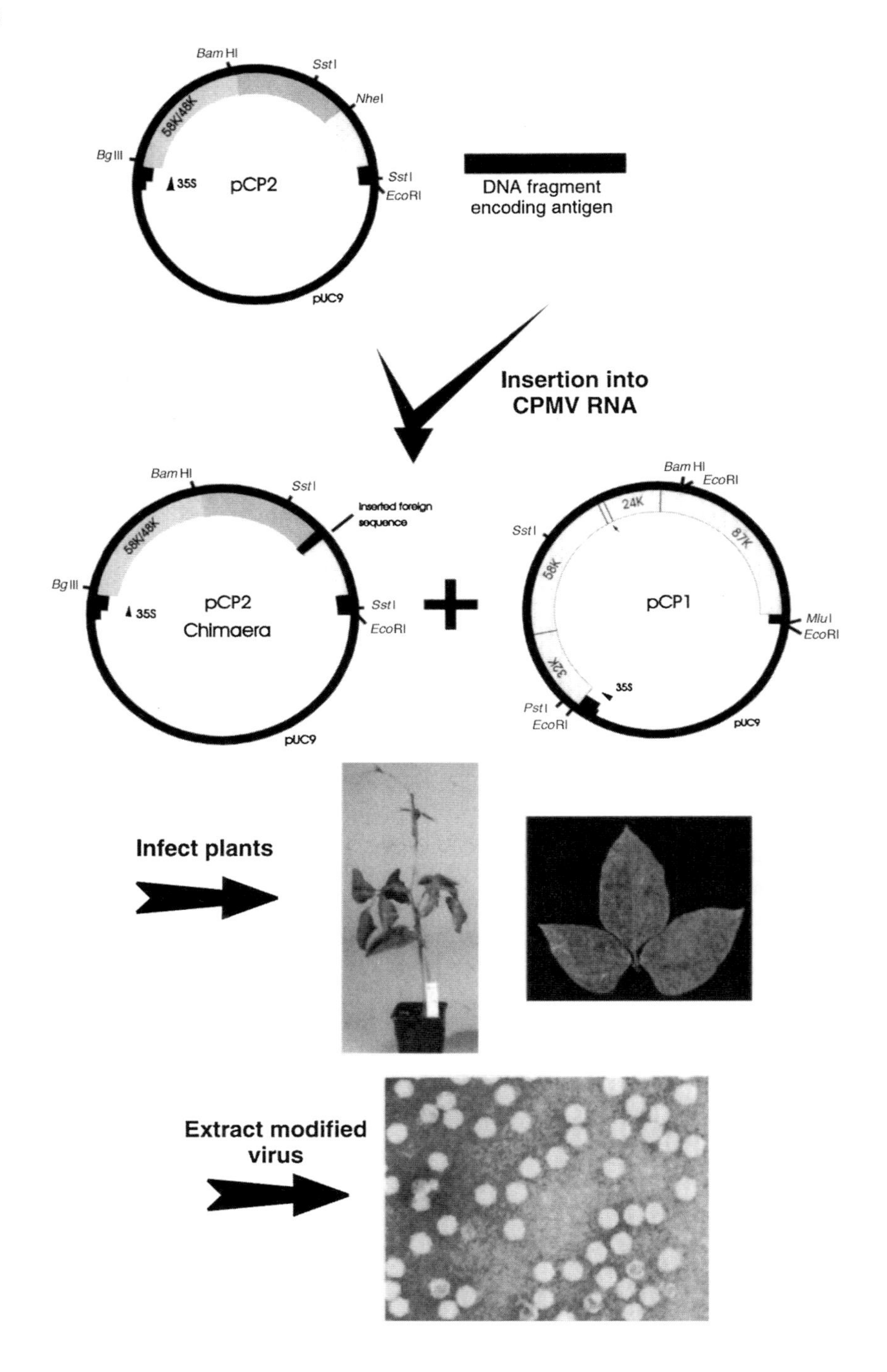

Scheme for the construction and propagation of CPMV-based chimaeras

[27] and appears to be common to most, if not all, chimaeras. Structural studies (see below) indicate that this cleavage did not result in the loss of the epitope from the surface of the virion but led to it being anchored at only its N-terminus.

To determine whether the modified S proteins of HRV-II and HIV-III had the antigenic properties of the inserted epitope, samples of the purified virions were analysed on Western blots, using either a polyclonal antiserum raised against HRV-14 or a monoclonal antibody against the HIV-1 epitope. In each case the modified S proteins reacted with the antisera but wild-type CPMV did not. The result confirmed that the denatured S proteins of the chimaeras retained the antigenic properties of the inserts. ELISA and immunogold-labelling studies on HIV-III particles confirmed that the inserted epitope was exposed on the surface of intact virions [28].

To assess the immunogenicity of HRV-II, samples of the purified chimaera or wild-type CPMV were injected into rabbits. The antisera were collected and used to probe Western blots of denatured HRV-14 particles. A single band corresponding to VP1, the protein from which the epitope was derived, was detected using antiserum raised against HRV-II particles even at a serum dilution of 1:16000 (Figure 5). The antisera did not react with the other coat proteins of HRV-14 (VP2 and VP3) and when serum raised against wild-type CPMV was used to probe the blots no reaction with any protein was observed. In addition, antisera raised against HRV-II reacted with formaldehyde-fixed HRV-14 particles. The antisera, however, proved to be non-neutralizing.

Figure 4

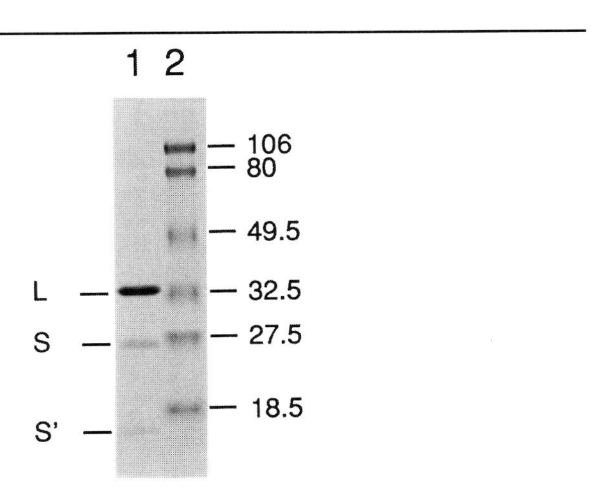

Coomassie-Blue-stained SDS-polyacrylamide gel of the coat proteins of HIV-III

Lane 1 contains SDS-denatured HIV-III virions with the positions of the L and modified S proteins indicated on the left. S′ is the form of the S protein that results from cleavage near the C-terminus of the insert (see text). Lane 2 contains prestained protein markers (Bio-Rad) with their molecular masses (kDa) indicated on the right.

Figure 5

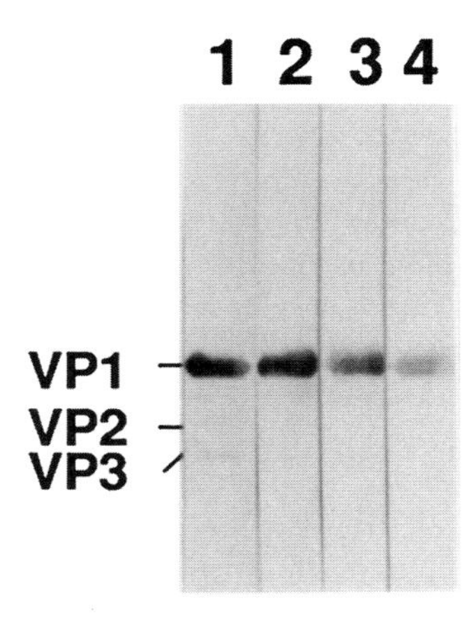

Immunogenicity of chimaera HRV-II

Samples of HRV-14 were denatured, electrophoresed on an SDS-polyacrylamide gel and the proteins transferred to a nitrocellulose membrane. The membrane was probed with serum raised in a rabbit that had been injected with the CPMV chimaera, HRV-II. The serum was diluted 1:2000 (lane 1), 1:4000 (lane 2), 1:8000 (lane 3) or 1:16000 (lane 4). The positions of the major structural proteins from HRV-14 (VP1, VP2 and VP3) are indicated on the left.

Purified HIV-III particles were injected subcutaneously into mice to determine the immunogenicity of the chimaera. Antisera obtained from the mice were found to bind to a peptide corresponding to residues 731–752 of HIV-1 gp41 in ELISA experiments [26]. The same antisera gave 97% neutralization of HIV-1 at a 1:100 dilution with a highly uniform response in all animals tested. Injection of as little as 1 μg of chimaera per mouse was sufficient to give up to 97% neutralization at a 1:200 serum dilution in three strains of mice [29]. The results compare very favourably with data compiled from neutralization studies carried out with epitopes presented using other expression systems. A bonus of the system was the unexpected observation that wild-type CPMV could elicit an anti-HIV-1 neutralizing antibody response, although this amounted to no more than 10% of that obtained with HIV-III. The neutralizing activity induced by wild-type CPMV, but not that induced by HIV-III, could be removed by adsorption with purified wild-type CPMV particles [26].

Structural analysis of the HRV chimaera

As the conformation of an epitope is vital in eliciting a therapeutically relevant immune response, knowledge of the precise configuration of an expressed epitope would be of great benefit for engineering potential vaccines. We have, therefore, solved the structure of HRV-II to 2.8 Å [27]. Since the atomic resolution structure of HRV-14 is known [30] it has been possible to determine whether the NIm-1A

Figure 6

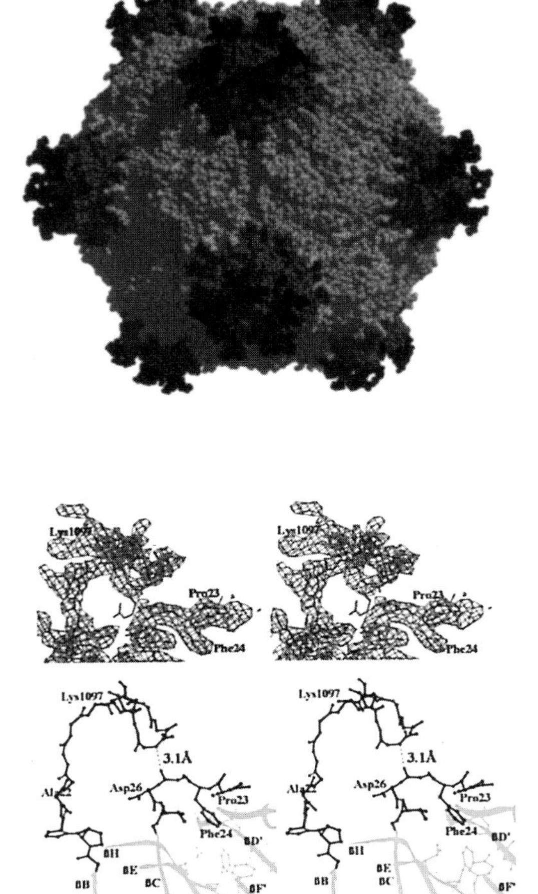

Presentation of NIm-1A site on the surface of CPMV in chimaera HRV-II

Top: space-filling model of HRV-II particles. The L subunit is shown in light grey, the S subunit in dark grey and the inserted epitope in black. Bottom: electron density for the chimaeric particle. The main chain of the modified βB-βC loop is clearly defined in the averaged electron density map and side chains of 12 of the 14 inserted residues were unambiguously modelled. The model is shown below the density map where the β-strands and side chains that comprise the immediate environment of Phe-24 are shown in light grey. Reproduced from [27] with permission.

site presented on CPMV adopts a configuration similar to that of the epitope in its native environment.

Crystals of HRV-II, measuring 1–1.5 mm in diameter, were prepared under conditions similar to those used for the preparation of crystals of wild-type CPMV [19,28]. X-ray diffraction patterns were obtained with measurable reflections beyond 1.9 Å [28], allowing a high-resolution electron density map to be calculated. Since the crystals of HRV-II had the same space group (I23) as those of wild-type CPMV [19], difference Fourier techniques could be used to solve the structure of the insert. Using this technique, all significant differences in density were found in the immediate vicinity of the βB-βC loop of the S protein, indicating that the rest of the CPMV structure remained undisturbed by the insertion of the heterologous sequence. A real space averaging procedure [31] was used to refine the structure. Electron density around the insertion site was weaker than that for residues not close to the βB-βC loop, indicating that the loop was somewhat flexible. However, the electron density was sufficiently above background to allow the polypeptide backbone to be traced precisely (Figure 6).

Thirteen of the HRV-14-specific residues formed a pseudo-closed loop on the surface of the chimaera, held by a hydrogen bond. The cleavage event occurring between the last two residues of the insert was evident in the structure of the chimaera, residues immediately following the cleavage site being clearly defined in the electron density map. The cleavage released the penultimate residue of the insert, allowing mobility and enabling the loop to attain a relatively smooth conformation with a separation at the base of 19 Å. In its native conformation, the NIm-1A epitope adopted a conformation including three sharp turns held by intra-loop hydrogen bonds. This conformation results from the constraint imposed by the short distance (9 Å) between residues at the beginning and the end of the sequence. If the residues of the loop on CPMV could be constricted to this relatively short distance apart, it is expected that the loop would adopt a conformation similar to that within the HRV-14 particle. The conformational differences between the NIm-1A site as expressed on the surface of CPMV and in its native environment probably explain why antibodies raised against the CPMV chimaera, though capable of binding to VP1, were non-neutralizing.

An analysis of the structure of the HRV-II chimaera showed that the CPMV-specific hydrophobic residue Phe-24 was buried within native CPMV residues near the base of the loop (Figure 6). This is an energetically favourable environment for this residue, allowed by cleavage of the loop. A hypothetical, uncleaved chimaera model was built in which the inserted residues would adopt the desired NIm-1A conformation. This model placed Phe-24 in an unfavourable environment, surrounded by hydrophilic residues. It is quite likely that the placement of Phe-24 in a hydrophilic environment encourages cleavage of the insert at the preceding Lys residue (Lys-1097), the penultimate amino acid of the HRV-14-specific epitope (Figures 2 and 6). If the residue at position 24 were to be changed from Phe to Arg, for example, the environment in that region may be improved, preventing cleavage. In addition, a mutation from Phe to Arg may allow the formation of a network of hydrogen bonds that would hold the two ends of the loop close together, thereby mimicking the native HRV-14 loop more closely. A construct with this mutation is being prepared for immunogenicity and structural studies. Crystallographic studies on other chimaeras, including HIV-III, are also currently underway.

Conclusion

This paper has emphasized the validity of antigen expression on the surface of CPMV for vaccine production. The development of a one-step cloning method for the introduction of antigenic sequences into the S protein of CPMV allows the straightforward construction of large numbers of chimaeras in order to assess their properties. One of the great advantages of the system we have described is the ease of propagation and purification of large quantities of chimaeras. These are genetically stable and display the antigenic properties of the insert. The prediction that the βB-βC loop of the S protein would provide a suitably exposed site for insertions has been proven by both immunogenicity studies and the structural data on HRV-II. However, other sites on CPMV are being investigated as potential sites for insertion of foreign sequences and early indications suggest that these chimaeras are also easy to propagate and purify. This opens up the possibility of simultaneously presenting more than one antigen on the surface of the virus.

The results from immunogenicity studies on HIV-III have been very encouraging and there are plans to extend the analysis. Recent work with transgenic potato plants expressing the *Escherichia coli* heat-labile enterotoxin [32] has indicated that it may be possible to use plant material directly as an edible vaccine. The fact that CPMV infects an edible plant, cowpea, means that the possibility of using chimaera-infected plant tissue for oral immunization can be investigated.

The ability to crystallize CPMV chimaeras and study the precise conformation of the epitope is a distinct advantage for this expression system. Studies on the structure of the HRV-II chimaera have already suggested how the chimaera could be modified so that the conformation of the epitope more closely resembles its native conformation. In the future, changes to the three-dimensional structure of the inserts will be correlated with variations in the immunogenicity of the expressed peptides so that the system can be optimized.

References

1. Francis, M.J. (1991) in Vaccines (Gregoriadis, G., Allison, A.C. and Poste, G., eds.), pp. 13–23, Plenum Press, New York
2. Schaaper, M.M., Lankhof, H., Pujik, W.C. and Meloen, R.H. (1989) Mol. Immunol. **26**, 81–85
3. Lomonossoff, G.P. and Johnson, J.E. (1995) Semin. Virol. **6**, 257–267
4. Lomonossoff, G.P. and Johnson, J.E. (1996) Curr. Opin. Struct. Biol. **6**, 176–182
5. Smith, G.P. (1985) Science **228**, 1315–1317
6. Greenwood, J., Willis, A.E. and Perham, R.N. (1991) J. Mol. Biol. **220**, 821–827
7. Mastico, R.A., Talbot, S.J. and Stockley, P.G. (1993) J. Gen. Virol. **74**, 541–548
8. Burke, K.L., Dunn, G., Ferguson, M., Minor, P. and Almond, J.W. (1988) Nature (London) **332**, 81–82
9. Dedieu, J.-F., Ronco, J., van der Werf, S., Hogle, J.M., Henin, Y. and Girard, M. (1992) J. Virol. **166**, 3161–3167
10. Arnold, G.F., Resnick, D.A., Li, Y., Zhang, A., Smith, A.D., Geisler, S.C., Jacobo-Molina, A., Lee, W.-M., Webster, R.G. and Arnold, E. (1994) Virology **198**, 703–708
11. Usha, R., Rohll, J.B., Spall, V.E., Shanks, M., Maule, A.J., Johnson, J.E. and Lomonossoff, G.P. (1993) Virology **197**, 366–374
12. Porta, C., Spall, V.E., Loveland, J., Johnson, J.E., Barker, P.J. and Lomonossoff, G.P. (1994) Virology **202**, 949–955
13. Turpen, T.H., Reinl, S.J., Charonenvit, Y., Hoffman, S.L., Fallarme, V. and Grill, L.K. (1995) Biotechnology **13**, 53–57
14. Fitchen, J., Beachy, R.N. and Hein, M.B. (1995) Vaccine **13**, 1051–1057

15. Evans, D.J. and Almond, J.W. (1991) Methods. Enzymol. **203**, 386–400
16. Lomonossoff, G.P. and Shanks, M. (1983) EMBO J. **2**, 2253–2258
17. van Wezenbeek, P., Verver, J., Harmsen, J., Vos, P. and van Kammen, A. (1983) EMBO J. **2**, 941–946
18. Goldbach, R. and van Kammen, A. (1985) in Molecular Plant Virology (Davies, J.W., ed.), Vol. **II**, pp. 83–120, CRC Press, Boca Raton, FL
19. Stauffacher, C.V., Usha, R., Harrington, M., Schmidt, T., Hosur, M. and Johnson, J.E. (1987) in Crystallography in Molecular Biology (Moras, D., Drenth, J., Strandberg, B., Suck, D. and Wilson, K., eds.), pp. 293–308, Plenum Press, New York
20. Chen, Z., Stauffacher, C.V. and Johnson, J.E. (1990) Semin. Virol. **1**, 453–466
21. Lomonossoff, G.P. and Johnson, J.E. (1991) Prog. Biophys. Mol. Biol. **55**, 107–137
22. Dessens, J.T. and Lomonossoff, G.P. (1993) J. Gen. Virol. **74**, 889–892
23. Klootwijk, J., Klein, I., Zabel, P. and van Kammen, A. (1977) Cell **11**, 73–82
24. Sherry, B., Mosser, A.G., Colonno, R.J. and Rueckert, R.R. (1986) J. Virol. **57**, 246–257
25. Kennedy, R.C., Henkel, R.D., Pauletti, D., Allan, J.S., Lee, T.H., Essex, M. and Dreesman, G.R. (1986) Science **231**, 1556–1559
26. McLain, L., Porta, C., Lomonossoff, G.P., Durrani, Z. and Dimmock, N.J. (1995) AIDS Res. Hum. Retroviruses **11**, 327–334
27. Lin, T., Porta, C., Lomonossoff, G.P. and Johnson, J.E. (1996) Folding Design **1**, 179–187
28. Porta, C., Spall, V.E., Lin, T., Johnson, J.E. and Lomonossoff, G.P. (1996) Intervirology, **39**, 79–84
29. McLain, L., Durrani, Z., Wisniewski, L.A., Porta, C., Lomonossoff, G.P. and Dimmock, N.J. (1996) Vaccine **14**, 799–810
30. Rossmann, M.G., Arnold, E., Erickson, J.W., Frankenberger, E.A., Griffith, J.P., Hecht, H.J., Johnson, J.E., Kamer, G., Mosser, A.G., Rueckert, R.R., Sherry, B. and Vriend, G. (1985) Nature (London) **317**, 145–153
31. Kleywegt, G.J. and Jones, A.T. (1994) in From First Map to Final Model (Bailey, S., Hubbard, R. and Waller, D., eds.), pp. 59–66, SERC Daresbury Laboratory, Daresbury
32. Haq, T.A., Mason, H.S., Clements, J.D. and Arntzen, C.J. (1995) Science **268**, 714–716

Oleosins as carriers for foreign protein in plant seeds

Maurice M. Moloney
Department of Biological Sciences, The University of Calgary, 2500 University Drive N.W., Calgary, Alberta T2N 1N4, Canada

Introduction

Most plant seeds store part of the energy needed for germination in small organelles called oilbodies. Oilbodies are, typically, spherical structures, comprising an oil droplet of neutral lipid (most frequently triacylglycerols) surrounded by a half-unit phospholipid membrane. The surface of these oilbodies is normally covered with a unique class of proteins called oleosins. These proteins have an extremely hydrophobic core, making them the most lipophilic proteins reported in databases. Their N- and C-termini are more hydrophilic or amphipathic [1]. Oleosins become associated with nascent oilbodies during oleosome biogenesis on the endoplasmic reticulum (ER) by a co- or post-translation mechanism [2,3]. Oleosins in oilseeds accumulate at relatively high levels. In the *Brassica* species, for example, oleosins may make up somewhere between 8 and 20% of total seed proteins [4]. This level of accumulation implies that oleosin genes are quite strongly transcribed during seed development.

When oilseeds containing oleosins are extracted in aqueous solvents they form a three-phase mixture of insoluble material, an aqueous extract and an emulsion of oilbodies that on standing or low-speed centrifugation will result in the floatation of the oilbodies replete with their oleosin complement. Successive aqueous washes of this oilbody fraction remove any extraneous surface proteins from the oilbodies. The oleosins, conversely, remain tightly associated with the oilbodies owing to their highly lipophilic core. Protein analysis of oilbody preparations that have undergone only floatation, centrifugation and washing reveals that virtually all the other seed proteins are absent, thus the oleosin fraction is highly enriched (Figure 1).

This observation led to the idea that oleosins could be excellent vehicles or carriers for heterologous proteins expressed in plant seeds. The utility of this would be the expression and simple purification of recombinant proteins in plants. As it has been possible to express heterologous proteins in plants for many years, it is interesting to ask why plants have not been used widely for recombinant protein production. The reasons do not seem to be related to the basic cost of plant production as on a cost-per-kilogram basis plants are a very inexpensive source of protein. It appears that the major limitation relates to the cost of extraction and purification. If oilbodies could be used as carriers it could

Figure 1

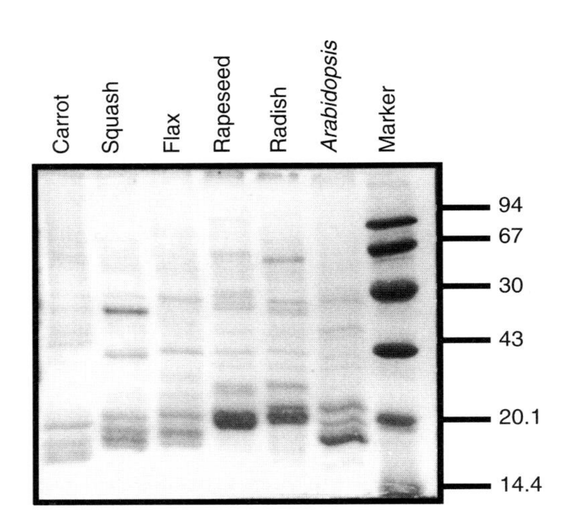

Coomassie-Blue-stained SDS/PAGE gel showing the oilbody fractions from a variety of common dicotyledonous seeds

The oilbody proteins were enriched only by one round of floatation centrifugation as discussed in the text. This centrifugation separates the rest of the cellular proteins from the oilbody proteins or oleosins. Fusions with these oleosins may be used as a vehicle for the subcellular targeting and subsequent purification of desired recombinant proteins in seeds.

impact favourably upon the economics of production of recombinant proteins in plants.

Recombinant oleosins and their subcellular targeting

While native oleosins are targeted with high avidity to oleosomes *in vivo*, it was not clear whether the addition of polypeptide sequences to oleosins would result in aberrant targeting. Such aberrant targeting has been noted previously with modified storage proteins such as recombinant phaseolin [5] or 2S albumin [6]. Experiments were performed to test the idea that modifications of oleosins at either the N- or C-terminal end might affect overall targeting to oilbodies [7,8]. These experiments showed that both N- and C-terminal translational fusions of oleosin with β-glucuronidase (GUS), did not significantly impair the basic targeting mechanism. The C-terminal GUS fusion remained enzymically active and this activity was followed in oilbody extracts to test whether the oleosin–GUS protein was attached to the oilbody with similar avidity to that of native oleosin. No differences in avidity could be detected. These experiments demonstrated a number of important properties of oleosins. They can be extended at either end and will still undergo oilbody targeting. Long polypeptide extensions do not seem to pose a problem (GUS has a M_r of 67000). Furthermore, these experiments were performed using a GUS known to be susceptible to N-glycosylation at position

358. It has been shown that if such a GUS is exposed to the ER lumen, GUS is inactivated owing to glycosylation [9]. In the case of the C-terminal extensions, GUS was fully functional, suggesting that it was not glycosylated. This indicates that fusions at the C-terminus of the oleosin are not exposed to the ER lumen and remain on the cytoplasmic side.

One further refinement to this subcellular targeting methodology was to interpose a specific labile cleavage site between the oleosin and recombinant protein domains. In our early experiments these were four amino acid proteolytic sites for enzymes such as thrombin or Factor Xa. Such a configuration should permit the recombinant polypeptide domain to be cleaved from the surface of the oilbodies. (See Figure 2). As can be seen from Figure 3 this cleavage can indeed be effected using oilbodies suspended in cleavage buffer and subjected to a specific protease treatment.

Application to production of the anticoagulant, hirudin

While the above experiments demonstrated the basic principles of oilbody-based recombinant protein production, it is of great interest to apply it to cases where the recombinant protein is of high value in order to test the economics as well as the technical aspects of the proposed system. To this end, we created a translational fusion between an *Arabidopsis* oleosin coding sequence and the coding sequence of the mature form of the blood anticoagulant, hirudin. Hirudin is a naturally occurring thrombin inhibitor produced and secreted in the salivary glands of the medicinal leech, *Hirudo medicinalis.* Hirudin is a highly effective anticoagulant with a number of very desirable properties, including stoichiometric inhibition of thrombin, short clearing time from the blood and low immunogenicity. The unit cost of production in leeches would be prohibitive for extensive therapeutic applications. Hirudin has been made in a variety of microorganisms, including *Escherichia coli* and yeast, but these entail significant fixed costs associated with fermentation. Our objective, therefore, was to test the oleosin expression and purification system as an alternative and potentially inexpensive source for hirudin.

Constructs containing a translational fusion of oleosin–hirudin under the transcriptional control of an oleosin promoter from *Arabidopsis* were introduced into *Brassica napus* using *Agrobacterium*-mediated transformation [10]. The resulting transgenic plants yielded seed in which a protein of M_r approx. 25 000 could be detected that cross-reacted with a monoclonal antibody raised against hirudin. This protein proved to be associated tightly with the oilbodies and could not be removed from the oilbodies by successive washings. Attempts to determine whether the oleosin–hirudin fusion protein had anti-thrombin activity suggested that the fusion protein was completely inactive. As part of the construction of the translational fusion, we had incorporated four additional codons specifying a Factor Xa cleavage site (I-E-G-R). This sequence was designed to permit release of the hirudin polypeptide sequence from oilbody by proteolytic cleavage. When the washed oilbodies from these seeds were treated with Factor Xa, hirudin polypeptide was released into the aqueous phase.

Figure 2

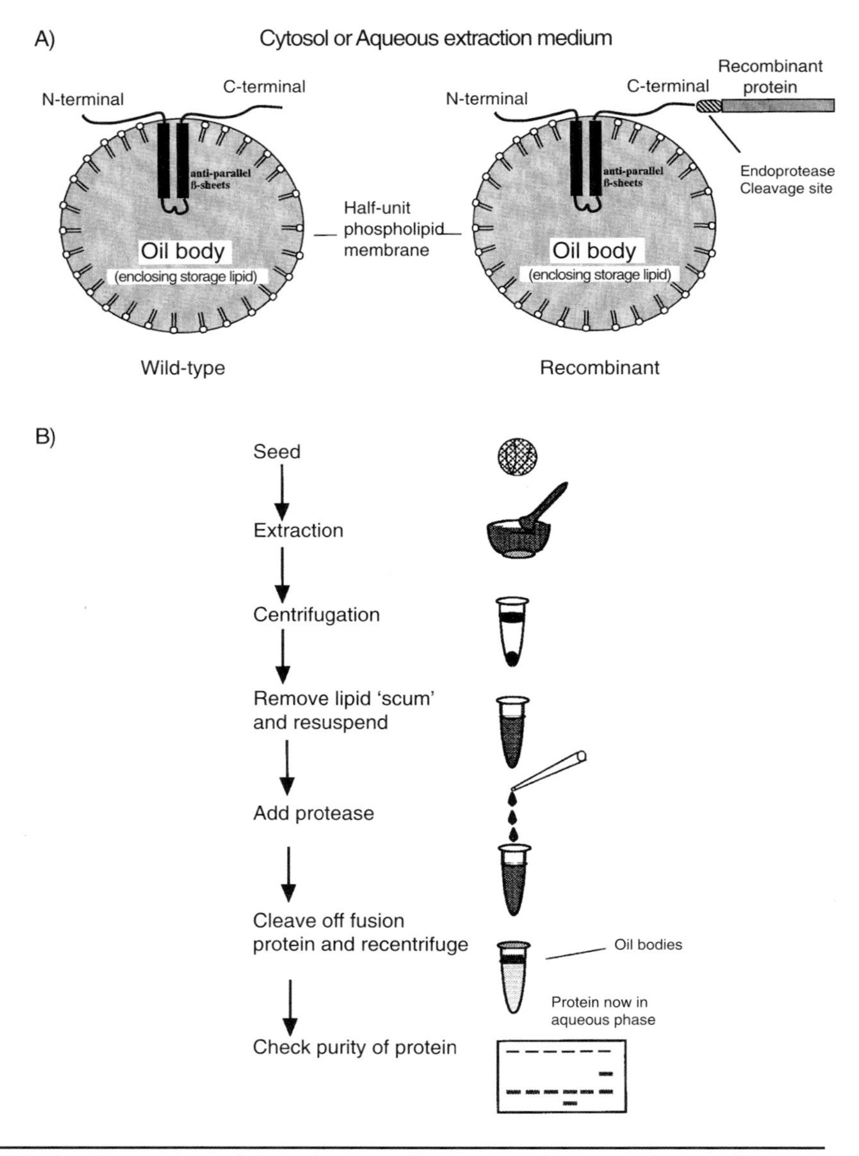

(A) Configuration of oleosins on wild-type and recombinant oilbodies based on models from Huang [5] and van Rooijen and Moloney [13]. (B) Flow diagram indicating extraction and purification steps for recombinant proteins expressed as oleosin fusions

Authentic hirudin has three disulphide bonds, which are essential to its activity. Thus, if the plant were to make hirudin polypeptide but did not allow its correct folding, no thrombin inhibition would be detected unless a refolding treatment was applied. In fact, after Factor Xa treatment, the aqueous phase showed strong anti-thrombin activity. The specific activity of the inhibitor was similar to that of

Figure 3

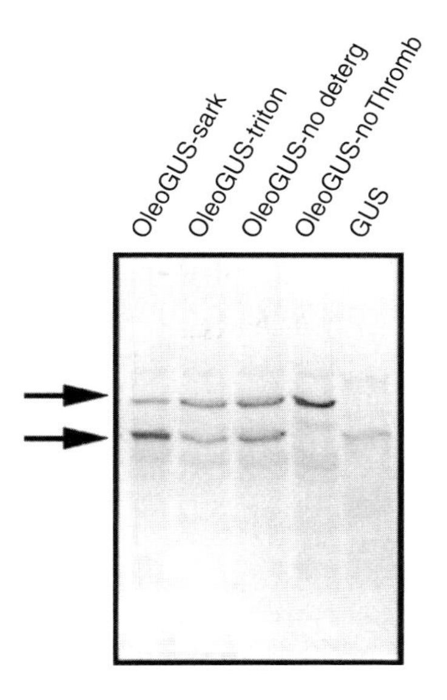

Western blot analysis of cleavage of oleosin–GUS and protein fusions from oilbodies of transgenic plants

*Oilbodies (5–10 μl) were re-suspended in 50 μl thrombin cleavage buffer made up of triethanolamine (0.1 M; pH 8.4). All oilbody samples, except for the controls (lanes 4 and 5; no thrombin and GUS) were treated with 0.02 IU thrombin. Oilbody samples were supplemented with no detergent (lane 3; no deterg.), 2% Triton X-100 (lane 2; triton) or 0.5% sarkosyl (lane 1; sark). Samples were incubated overnight at 37 °C and the reaction completed the next morning by raising the temperature to 55 °C for 30 min. All the protein samples were run on a 7.5% SDS/PAGE gel, blotted on PVDF membranes and visualized immunologically using GUS-antiserum (Clontech) as primary antibody with a goat anti-rabbit IgG linked to alkaline phosphatase as secondary. The arrows indicate the relative molecular mass of full-length oleosin–GUS (M_r = 87670) and the cleaved/native GUS (M_r = 68200). Lane 5 (GUS) comprises a total soluble protein sample (100000 × **g** supernatant) from a plant expressing a transcriptional fusion of GUS. This acts as a molecular mass marker and control for plants expressing untargeted GUS.*

recombinant hirudin secreted from yeast cells, indicating that the majority of the hirudin released was correctly folded and with the correct disulphide bonds. The released hirudin was subjected to ion-exchange chromatography (Mono Q). The fractions showing anti-thrombin activity were then concentrated and loaded on to a C18 reversed-phase analytical HPLC column. This showed that the hirudin was substantially pure after the ion-exchange step. The HPLC trace showed two peaks close to the expected retention volume of hirudin. Mass spectrometric analyses of these two peaks using time-of-flight–matrix-associated laser desorption ionization revealed that the two peaks were full-length hirudin and a truncated product from which two C-terminal amino acids were missing. Interestingly, such a truncated form was also found when hirudin was expressed in yeast [11].

A number of improvements would be required to render this system economical. These are in (a) the expression levels, (b) cleavage efficiency and (c) cleavage system (as Factor Xa would be prohibitively expensive). Expression levels of the recombinant protein have in a few cases been about one-tenth of the native oleosin accumulation. This corresponds to about 1% of total cellular protein. It is, however, expected that, through the use of stronger promoters than the native oleosin promoters, higher levels of accumulation will be sustained. Cleavage efficiency varies from one fusion protein to another and may be affected by the conformation of the target protein and also by the spacing between the cleavage site and the oilbody surface. Finally, the use of Factor Xa here was essentially as a convenient experimental system. In practice it would be essential to have an inexpensive cleavage enzyme or chemical to render this process competitive. Recent experiments with inexpensive enzymes such as bacterial clostripain or collagenase suggest that alternative cleavage systems are compatible with this process and would allow the cost of the cleavage step to be minimized.

Other proteins or polypeptides expressed as oilbody fusions

The potential for using oleosin fusions as an adaptable means of recombinant protein expression in seeds is under investigation. Table 1 shows a number of examples that illustrate the range of applications that may be amenable to this technology. It is interesting to note that the production of oleosin–polypeptide fusions appears to function for rather short peptides such as interleukin 1β and hirudin as well as for much longer polypeptides such as GUS. In cases where larger polypeptides are produced there is good evidence that proper folding occurs. This is shown by the fact that in a number of cases, including GUS and xylanase, the fusion protein retains its enzymic activity. This finding has also led to experiments that illustrate the utility of oilbodies as immobilization matrices [12,13]. Using GUS as a model it was shown that a dispersion of oilbodies carrying GUS on their surface would hydrolyse glucuronide substrates for GUS with catalytic properties indistinguishable from soluble GUS. Furthermore, it was shown that virtually all the enzymic activity could be recovered and recycled by the use of floatation centrifugation to obtain the oilbodies. The enzyme in dry seed remains fully active for several years. Once extracted on oilbodies the half-life of the enzyme was about 4 weeks [7].

Conclusions and future prospects

The use of plant oilbodies and their associated proteins, oleosins, as vehicles for recombinant protein product has been illustrated with a number of examples. The major advantage of using oilbodies as carriers is the ease with which proteins can be recovered and purified. Oleosin targeting does not seem to be impaired even by very long polypeptide extensions to the N- or C-termini of the oleosin. This greatly enhances the versatility of this system in contrast to alternative approaches for recombinant protein production in plants. Separation of oilbodies from seed extracts is amenable to scale-up using equipment typical to dairy operations, such as cream-separators [14]. In consequence, it seems likely that this system could be

Table 1

Fusion protein (source)	M_r	Use
High-methionine Zein fragment (corn)	10000	Alteration of amino-acid content of oilseeds
Xylanase (*Neocallimastix patriciarum*, Rumen fungus)	68000	Production of xylanase for animal feed supplement or in wood pulping
Single-chain antibodies (synthetic, human/mouse)	30000	Inexpensive production of antigen binding proteins for diagnostic or bioimaging applications
Proteases (various sources)	30–50000	Cleavage enzymes, meat tenderizers, biological detergents
Hirudin (medicinal leech)	6000	Anticoagulant, thrombin inhibitor
Interleukin-1β (human)	~1000	Cytokine, anti-tumour
β-Glucuronidase (*Escherichia coli*)	~68000	Marker enzyme. Model for targeting experiments.

Proteins and polypeptides which have been expressed in seeds as oleosin fusions

used for the production of a wide range of proteins of therapeutic, industrial and feed or food use. We are investigating production of a wide range of commercially attractive polypeptides in this system and concurrently developing scale-up extraction and purification systems that could be applied to different oilseeds engineered for production of such oleosin–polypeptide fusions.

References

1. Huang, A.H.C. (1992) Oilbodies and oleosins in seeds. Annu. Rev. Plant Physiol. Plant Mol. Biol. **43**, 177–200
2. Hills, M.J., Watson, M.D. and Murphy, D.J. (1993) Targeting of oleosins to the oil bodies of oilseed rape (*Brassica napus L.*). Planta **189**, 24–29
3. Loer, D. and Herman, E.M. (1993) Cotranslational integration of soybean (Glycine max) oil body membrane protein oleosin into microsomal membranes. Plant Physiol. **101**, 993–998
4. Huang, A.H.C. (1996) Oleosins and oil bodies in seeds and other organs. Plant Physiol. **110**, 1055–1061
5. Hoffman, L.M., Donaldson, D.D. and Herman, E.M. (1988) A modified storage protein is synthesized, processed and degraded in the seeds of transgenic plants. Plant Mol. Biol. **11**, 717–729
6. Krebbers, E. and Vandekerckhove, J. (1990) Production of peptides in plant seeds. Trends Biotechnol. **8**, 1–3
7. Van Rooijen, G.J.H. and Moloney, M.M. (1995) Plant seed oil-bodies as carriers for foreign proteins. BioTechnology **13**, 72–77
8. Van Rooijen, G.J.H. and Moloney, M.M. (1995) Structural requirements of oleosin domains for subcellular targeting to the oil body. Plant Physiol. **109**, 1353–1361
9. Iturriaga, G., Jefferson, R.A. and Bevan, M.W. (1990) Endoplasmic reticulum targeting and glycosylation of hybrid proteins in transgenic tobacco. Plant Cell **1**, 381–390
10. Moloney, M.M., Walker, J.M. and Sharma, K.K. (1989) High efficiency transformation of *Brassica napus* using *Agrobacterium* vectors. Plant Cell Rep. **8**, 238–242
11. Heim, J., Takabayashi, K., Meyhack, B., Märki, W. and Pohlig, G. (1994) C-terminal proteolytic degradation of recombinant desulfato-hirudin and its mutants in the yeast *Saccharomyces cerevisiae.* Eur. J. Biochem. **226**, 341–353
12. Moloney, M.M. and Van Rooijen, G.J.H. (1996) Recombinant proteins via oleosin partitioning. Inform **7**, 107–113
13. Kühnel, B., Holbrook, L.A., Moloney, M.M. and Van Rooijen, G.J.H. (1996) Oil bodies of transgenic *Brassica napus* as a source of immobilized β-glucuronidase. J. Am. Oil Chem. Soc. **73**, 1533–1538
14. Jacks, T.J., Hensarling, T.P., Neucere, J.N., Yatsu, L.Y. and Barker, R.H. (1990) Isolation and physiocochemical characterization of the half-unit membranes of oilseed lipid bodies. J. Am. Oil Chem. Soc. **67**, 353–361

Rubber latex as an expression system for high-value proteins

H.Y. Yeang*‡, P. Arokiaraj*, Hafsah Jaafar*, Samsidar Hamzah*, S.M.A. Arif* and H. Jones†
*Biotechnology and Strategic Research Division, Rubber Research Institute of Malaysia, RRIM Experiment Station, 47000 Sungei Buloh, Selangor D.E., Malaysia and
†Division of Biosciences, University of Hertfordshire, Hatfield, Herts AL10 9AB, U.K.

Introduction

Natural rubber comes from the latex of the commercial rubber tree, *Hevea brasiliensis*, which is a member of the Euphorbiaceae. The tree originates from South America but is today cultivated mainly in South East Asia. Latex is synthesized in specialized cells, the laticifers or 'latex vessels', which are articulated tubes found in all parts of the rubber tree, from the roots to the trunk, branches, flowers and fruits. For the purpose of commercial exploitation, however, tapping of latex vessels is carried out only on the trunk where the laticifers are laid out in concentric sheets of laterally anastomosing vessels from the cambium to the periphery. In a transverse section of the trunk, the concentric sheets of laticifers appear as rings (much as xylem vessels of deciduous trees appear as annual rings). A review of *Hevea* bark and laticifer anatomy has been given by Gomez [1]. Tapping of the tree involves the excision of a shaving of bark on a sloping cut. The skill of the tapper enables him to cut deep into the bark to sever as many latex vessel rings as possible for maximum yield, but without wounding the cambium, which would lead to uneven bark regeneration. High turgor pressure (approx. 10 atm) within the latex vessels forces the latex out of the cut vessels [2,3]. Latex flow ceases after 1–3 h because a plugging mechanism gradually seals off the severed ends of the latex vessels [4]. Tapping on alternate or third days allows sufficient rest for the tree to regenerate the latex exuded.

The present state of the Malaysian rubber industry

The Malaysian natural rubber industry is in transition. After a peak production of 1.66 million tonnes of natural rubber in 1988 [5], Malaysia has since relinquished its position as the world's leading natural rubber producer. While the country's output of 1.09 million tonnes [6] still accounts for a sizable 19% of world production, various factors have contributed to the slow-down in the upstream rubber industry. Paramount among these is the swing of the economy from

* To whom correspondence should be addressed

primary production towards industrialization on the one hand, and to the shortage of estate labour, which the rapid upsurge of industrialization exacerbates, on the other. The hectareage under rubber cultivation was 2.1×10^6 ha in 1971 [7], but had declined to 1.7×10^6 ha by 1995 [6]. The harvesting of rubber by tapping is labour-intensive and has thus far defeated attempts at automation and mechanization. Skilled rubber tappers who can extract high yields without wounding the tree are now scarce, and this has resulted in vast tracts of mature rubber plantings being left untapped. The problem is likely to become more acute in the years to come.

Seeking non-rubber products from the rubber tree

The Malaysian natural rubber industry could perhaps be better viewed as the 'rubber tree industry' rather than the rubber industry. Besides rubber, which is the main commercial product of the rubber tree, the exploitation of non-rubber products from the whole tree needs to be evaluated. The most important commercial non-latex product from the rubber tree at the moment is rubberwood, a much sought after commodity in furniture manufacture [8]. From latex, there are also various non-rubber constituents that, with appropriate research and development, might be exploited profitably. Latex serum is naturally rich in proteins, sugars, lipids, carotenoids, inositols, organic acids, nucleic acids and minerals [9,10]. It is conceivable that, in time, various natural products might be recovered from the latex serum that is presently discarded during rubber processing. As an example, quebrachitol from latex serum is currently being explored for pharmaceutical uses [11]. Over and above the native natural products of the trees, the advent of genetic engineering techniques has today introduced new possibilities in their exploitation. The harvest from the rubber tree need no longer be limited by the confines of its genetic blueprint; products foreign to *Hevea* are within the bounds of possibility.

The Rubber Research Institute of Malaysia (RRIM) is looking into ways to upgrade productivity and profitability of rubber cultivation by exploiting various non-rubber products obtainable from the rubber tree. In this connection, the RRIM is actively investigating the insertion of genes, for the production of commercially valuable proteins, into the rubber tree. These genes are engineered to express themselves in the latex that can then be harvested by tapping the rubber tree [12]. Essentially then, each transgenic rubber tree becomes a mini-factory dedicated to the production of the target protein. It is intended that once the technique for gene insertion has been established, its exploitation would not be limited to any one product but could in theory be harnessed for a multitude of proteins or protein-based products for which the genes are available. The RRIM is hopeful that the project would lead to a novel and highly profitable system of synthesis of recombinant proteins such as pharmaceuticals. Apart from the prospect of profitability from the commercial viewpoint, the possibility exists that various forms of chemotherapy, hitherto prohibitively expensive, could be placed within reach of the man-on-the-street through increased efficiency in pharmaceutical production.

Rationale for producing foreign proteins in the rubber tree

Existing technology already enables recombinant proteins to be produced by transgenic bacteria, yeasts and filamentous fungi. So why venture on to new ground with higher plants? Micro-organisms need more than casual attention to thrive; they need intensive care. The ambient temperature, pH and aeration must be carefully controlled and kept just right. Nutrients must be added in carefully regulated doses and waste products removed. Micro-organisms are cultured in high-technology fermenters that are housed in expensive-to-run factories, which result in high manufacturing overhead costs. Not surprisingly, this all adds up to high costs for the end products. For example, the prescription-drug version of human tissue plasminogen activator (t-PA) produced in fermenters by Genentech reportedly sells for as much as US$22 000 per gram, partly to recover research costs [13].

Each transgenic plant that produces the target pharmaceutical is a factory (effectively an individual manufacturing plant) in itself, with the advantage that plants take care of themselves, requiring little more than sunlight, water and basic horticultural input. The potential savings in pharmaceutical manufacture by this cost-efficient approach can be very substantial indeed. An important advantage of plants over bacteria is that glycosylation of expressed eukaryote proteins would take place in the plant system, which does not occur with bacteria. Furthermore, plants as natural protein factories are solar-powered, rendering them ecologically friendly.

Thousands upon thousands of plant species could be candidates, but the rubber tree avoids many disadvantages inevitably associated with most of the others. For example, in most plants the target pharmaceutical can only be produced in small quantities in the plant tissue, so its extraction can be problematic and inefficient. With practically all other plants, continual harvesting is not possible, as recovery of the target product involves the harvesting and destruction of the plant or a portion of the plant. Where the harvested plant is not destroyed, it would require a considerable period of recovery for re-growth before a subsequent harvest is possible. This is where the rubber tree differs from the rest. *H. brasiliensis* produces voluminous sap (latex) upon tapping, which is a non-destructive method of latex extraction. Each mature tree produces some 70–150 ml or more of latex per tapping and the tree can be tapped every alternate day throughout the year without pause. Hence, the transgenic *Hevea* system allows for continual production of the target protein. Latex serum contains approx. 1.5% native proteins, and so the mechanisms for protein production are present. The copious latex output from the rubber tree is matched by the prodigious rate of latex regeneration that is facilitated by the fact that the nuclei of the laticifers are not generally expelled with the latex outflow [14,15]. Laticifers in mature *Hevea* bark are multinucleate [14] and the presence of the parietal nuclei suggests that the transcription of foreign proteins in transgenic *Hevea* could occur within the laticifers.

Besides the target high-value proteins, the by-products that can be obtained from the transgenic rubber tree are rubber and, eventually, cultivated tropical timber, both of which are valuable commodities in their own rights.

A parallel approach to recombinant protein production currently being evaluated has involved the use of transgenic animals. The successes in the *in vivo* production of human proteins in cattle and goat milk [16–20] give a strong signal that similar success with *Hevea* latex is a realistic goal that is well within reach. Even if the yields of recombinant protein turn out to be similar, there are extra advantages with the rubber tree; the costs of maintaining animals is obviously far higher than the costs of maintaining plants. Moreover, there is substantial opposition from some quarters to the commercial exploitation of transgenic animals.

Ease of propagation is another factor to consider, especially as the transgenic rubber system can exploit simple clonal expansion techniques. The amenability of *Hevea* to vegetative propagation means that the acquisition of sufficient numbers of transgenics to support a viable commercial undertaking would not be a problem. Although the rubber tree can be propagated directly from seed, almost all rubber trees presently cultivated in Malaysia are bud-grafted clones of superior cultivars. Hence, the method and infrastructure for vegetative propagation of the rubber plant by conventional horticultural approaches are well entrenched in the natural rubber industry [21]. With the transgenic rubber tree there is, therefore, no necessity to obtain large numbers of transgenic plants *de novo*, where the levels of gene expression in individuals might be variable and require testing. The first requirement with *Hevea* would be to select from a number of transformants those that give the best recombinant protein expression. From these, unlimited numbers of clonally identical plants can be multiplied vegetatively. Since tissue culture is not involved in the horticultural process, deleterious effects of somaclonal variation are not a potential problem.

Hevea latex as a carrier for recombinant proteins

Fresh natural rubber latex is a colloidal mixture consisting mainly of rubber particles suspended in an aqueous medium, called the C-serum. When fresh latex is centrifuged at high speed (44000 ***g***) it separates into three main zones: the rubber fraction at the top (since rubber has a density slightly less than 1), a heavy 'bottom fraction' and the C-serum in between. The bottom fraction is composed mainly of fluid-filled bodies called lutoids, but also contains other minor organelles [22].

Latex C-serum, which constitutes some 50% of the volume of latex, contains approx. 12 mg of protein$\cdot$ml^{-1} (H.Y. Yeang, unpublished work). There is a very wide variety of proteins present in C-serum; for example, all the glycolytic enzymes of the respiratory pathway are found there [23]. C-serum contains 2 to 10 mg of sugars$\cdot$ml^{-1} (H.Y. Yeang, unpublished work), mainly sucrose, and it is also rich in cyclitols, especially quebrachitol [24]. The rubber particles make up approx. one-third of the volume of latex. Rubber particles are spherical or ovoid and are composed mainly of *cis*-polyisoprene (i.e. natural rubber); most of them are very small, with a mean diameter of approx. 0.07 μm. A small proportion of rubber particles are larger, with diameters as large as 1 μm or greater, and these account for the bulk of the rubber in latex [25]. The rubber particle is surrounded by a single protein–lipid membrane, the protein moiety of which is insoluble. Besides rubber particles, the other major organelle in natural rubber latex is the lutoid, which take up 15–20% of the latex volume. Lutoids are vacuole or lysosome-like vesicles bound by a single membrane and are 0.5–3 μm in diameter [14]. The fluid found in lutoids, the B-serum, also contains proteins and sugars. Latex B-serum contains approx. 24 mg of soluble protein·ml^{-1} (H.Y. Yeang, unpublished work), 50–70% of which is hevein [10,26]. There are approx. 4 mg of B-serum proteins in 1 ml of natural rubber latex.

Disregarding the insoluble proteins on the surfaces of the rubber particle and on the lutoid membranes, approx. 10 mg of soluble proteins are present in 1 ml of latex. This gives some idea of the protein-carrying capacity of natural rubber latex. A mature rubber tree of the clone Gl 1 (not a high-yielding cultivar) that is used in the RRIM transformation experiments (see below) produces approx. 70 ml of latex per tapping. A tree therefore produces 700 mg of soluble proteins per tapping. Taking planting density to be 400 trees per hectare and tapping to be carried out on alternate days, some 50 kg of soluble native latex proteins are produced in the latex per hectare per year. As latex output can be stimulated by the application of ethephon (2-chloroethyl phosphonic acid) [27,28], the yield may be increased a further 20–40% with judicious ethephon treatment, as is commonly practised in rubber cultivation.

The rubber tree normally comes into tapping after approx. five years, when the trunk has reached a girth of 50 cm. Opening of the tree at too early an age could cause its further growth to be retarded because of the partition of assimilates between the exuded latex and tree growth. Nevertheless, staggered commencement of tapping of the immature trees may take place at two or three years of age, should the economic return make this worthwhile. This would be the case if the value of the recombinant protein obtained justifies the resulting decrement in latex output.

As mentioned above, the advantage of using transgenic *Hevea* for recombinant protein production lies essentially in targeting the protein expression in the latex as this would allow for continual, non-destructive harvesting. It would be best for the protein to be expressed in the C-serum as this is the most abundant aqueous constituent of natural rubber latex and it can be recovered readily by centrifugation. Nevertheless, as latex is a complex fluid containing diverse organic and inorganic constituents [9], post-harvest purification of the recombinant protein will be necessary.

Prospects for commercializing transgenic *Hevea*

The present day high cost of many recombinant protein-based pharmaceuticals is the reason that bioreactor production of the proteins can be sustained commercially. Protein-based pharmaceuticals in general, and especially those occupying niche markets, are of high value while they are produced in relatively low volume. The transgenic *Hevea* system can be expected to compete favourably with bioreactors within these parameters, for its inherent advantages lie in both the low costs required in its basic maintenance and the ease of production scale-up. Taking advantage of the conventional horticultural methods available for vegetative propagation, a scale-up to hectares of genetically identical clones can be undertaken with minimal fuss and at modest cost. Hence, it is envisaged that the transgenic rubber tree would be particularly advantageous in the high-volume production of proteins of moderate value, where bioreactor production would be less economically attractive. Examples of such proteins include those that might be used in mass market consumer products.

Whether the transgenic rubber tree can become a successful manufacturing plant will depend firstly on whether the inserted gene can be effectively transcribed and translated in the latex. As described below, the system has shown promising results with the marker gene *gus*, which codes for the enzyme β-glucuronidase (EC 3.2.1.31). Experiments with other genes are ongoing. To some extent, success could hinge upon the opportune insertion of the foreign gene at a site in the host genome that allows the gene to be strongly expressed. But, as mentioned above, the transgenic *Hevea* system does not depend upon repeated success with genetic transformation, because clonal copies can be vegetatively propagated from just a handful of selected high-expression transformants. Several other crucial factors would determine its ultimate commercial viability. The rate of synthesis of the recombinant protein is one such major consideration. Gene expression would need to be high enough to be competitive with alternative manufacturing processes, especially conventional bioreactor production. As functional proteins are three-dimensional molecules with complex folding characteristics, it is not known with certainty if the transgenic rubber tree can correctly process for secretion into the latex all heterologous proteins of interest. It is also important that post-translation cleavage, if it occurs in the plant, is not flawed. Proteins may be glycosylated and, in some instances, may owe specific properties to the glycosylated state. As glycosylated latex proteins occur in latex, the essential apparatus for protein glycosylation – absent in bacteria – is already present in the laticifer. It is nevertheless important to determine if glycosylation of the recombinant protein is comparable with that of the native protein it seeks to duplicate.

The latex that flows out of the rubber tree is free of animal viruses (J.B. Gomez, personal communication). Recombinant proteins produced by the transgenic rubber tree are therefore safe from the danger of contamination from the AIDS virus, hepatitis virus and other pathogens that may be present in products generated from human blood, such as α-1-antitrypsin (AAT) and blood clotting factors.

In terms of initial start-up, the time-lag of the approx. three years immaturity period of the rubber plant before the first product can be harvested may not compare favourably with the bioreactor. The long reproductive cycle of the rubber tree (approx. five years to flowering) would also mean that manipulation of the inserted gene by sexual reproduction (e.g. to transfer foreign genes from two separately transformed plants into their progeny) [29] could be time-consuming. In compensation, the perennial nature of the rubber tree has its advantages. The tree has an economic life of approx. 30 years. Once the transgenic rubber tree is in production, maintenance costs are low, especially in comparison with running a bioreactor. Exact costs vary with different rubber growers but, generally, around US$840 is spent on field maintenance and latex production per hectare of mature rubber per year. This includes the costs of management and supervision, field sanitation and fertilizers, transportation, tapping and latex collection etc., but excludes the cost of processing the latex.

Field-release and biosafety issues

Biosafety aspects of transgenics are important considerations that inevitably invoke deliberation and controversy. Compared with many other crop plants, transgenic *Hevea* is probably placed in fewer equivocal positions with respect to field-release and biosafety when planted in Malaysia. Unlike many other species where the attributes of the plant itself are modified (e.g. the conferring of disease, herbicide or stress resistance), the immediate objective of *Hevea* genetic transformation does not so much seek to modify the plant as to use it as a vehicle for foreign protein production. The genes that have been inserted into the transgenic plant will, in most cases, have no significant effect on the metabolic functions of the rubber plant, nor of plants in general, so the genetic modification does not increase the biological competitiveness of the host. Hence, the possible impact of any gene escape to the ecosystem is expected to be small.

In the first instance, the containment of foreign genes inserted into transgenic *Hevea* will be less of a problem than with the same genes in most other plants. Since the rubber tree is not native to Malaysia, there are no wild *Hevea* in the country to which foreign genes might inadvertently spread. *Hevea* is not known to hybridize with any native genus in Malaysia. The fact that cultivated *Hevea* is propagated vegetatively, rather than directly from seeds, further distances any threat that escaped genes might have to the natural rubber industry. The rubber tree does not flower until it is approx. five years old, thereby providing a considerable time period for testing before it becomes sexually mature. The limit of *Hevea* pollen dispersal has been determined to be approximately 1.1 km [30]. Using this information, initial field release of transgenic *Hevea* can be restricted to isolated gardens away from any rubber cultivation while biosafety evaluation is underway.

The tapped latex from the transgenic rubber tree would require post-harvest purification to obtain the recombinant protein. Hence, the inserted gene is not presented to the consumer, unlike in the case of, for example, an unprocessed food product from a transgenic plant. The perception of potential danger posed by the presence of foreign genes in the product derived from transgenic organisms would not, therefore, be a critical issue with *Hevea.*

The *gus* gene as a model system for *Hevea* transformation

Genetic transformation of *H. brasiliensis* by *Agrobacterium,* as evidenced by gall formation and octopine synthesis in the host plant and by *in vitro* tumour growth on phytohormone-free medium, has been carried out by Arokiaraj and Wan Abdul Rahaman [31]. In this instance, the transfer of the unmodified (wild-type) *Agrobacterium* T-DNA was localized to the *Agrobacterium*-infected tissue. *Hevea* transformation involving engineered T-DNA that incorporated a foreign gene was first reported by Kitayama and co-workers [32]. Particle bombardment was used to insert a genetic construct, containing the *gus* cDNA linked to the cauliflower mosaic virus (CaMV) 35S promoter, into *Hevea* callus tissue. The first whole transgenic rubber plants were obtained by Arokiaraj and co-workers, who inserted the *gus* gene (pBI221) (see original work by Jefferson [33]) into anther callus tissue using the particle gun [34] and subsequently regenerated a plantlet through somatic embryogenesis. Some of the subsequent genetic transformations were carried out by co-cultivation with *Agrobacterium* GV2260 (p35SGUSINT), a vector system used previously by Vancanneyt et al. [35].

In *Agrobacterium*-mediated transformation [36], transgenic callus tissue (derived from anther callus) expressing the inserted gene for kanamycin resistance *(nptII)* and the marker gene *(gus)* were maintained on culture medium containing kanamycin. Kanamycin-resistant calluses were generated following the method of Chen [37]. Following subculture of transgenic calluses into differentiating medium, embryoids were obtained and three of these were regenerated into transgenic plants that were successfully established in soil. The presence of the *gus* gene in leaf tissue has been detected by Southern blotting analysis. GUS expression was demonstrated using 5-bromo-4-chloro-3-indoyl β-D-glucuronide (x-gluc) as the enzymic substrate. The GUS protein was detected in the leaf of the transgenic plant and, crucially, also in the latex obtained from the plant. In fact, histochemical examination showed that expression of the inserted gene was especially enhanced in the latex within the laticifers of the transformed plant. The CaMV 35S promoter has been reported to direct gene expression strongly to the phloem of transgenic tissue [38]. Therefore, the marked GUS expression in the laticifers is not surprising, as laticifers may be regarded as specialized phloem elements. Analysis of GUS expression in the fractions of latex obtained by centrifugation showed the GUS protein to be in the C-serum.

The transgenic rubber plants have been multiplied by bud-grafting using buds obtained from two of the transgenic plants. A total of 194 plants, representing four successive vegetative generations of genetic clones, were obtained subsequently, and every one of these has been shown to be positive for the GUS protein in the latex. GUS expression appears, therefore, to be stable, and neither reversion nor chimerism have been observed over the three vegetative generations. These results indicate that foreign genes can be expressed in the latex of the transformed rubber plant and that a single transgenic plant selected for high protein expression can be easily multiplied into any number of vegetative clones that retain the ability to express the protein.

Conclusions

In conclusion, the salient features of the exploitation of the transgenic rubber tree for foreign protein production can be summarized as follows:

The concept is a novel approach to cost-efficient recombinant protein production with the target product being produced in the latex of rubber trees that essentially serve as production lines. Application of the methodology, once developed, is not limited to any specific protein but is generally applicable to many protein or peptide-based products. The transgenic *Hevea* system may be especially well-suited to the high-volume production of moderately priced proteins.

The approach is environmentally friendly. The process is driven by the sun and is therefore energy-efficient and essentially pollution-free.

Rubber trees require no special attention beyond routine horticultural maintenance. Their use is thus highly cost-efficient.

Production of the target protein is continual through a system of non-destructive harvesting (tapping) of the latex of the rubber tree. Recovery of the product from the aqueous phase of the latex is simple and efficient.

Glycosylation of eukaryote proteins, which does not occur in bacterial systems of protein production, can take place in the transgenic rubber tree.

The latex that exudes from the rubber tree is free of animal viruses.

Successful transformation of the rubber tree for a specific gene needs to be achieved only once. Rubber trees are amenable to clonal propagation and an unlimited number of genetically identical plants (clones) can be generated by conventional horticultural methods.

The technology does not involve the use of animals and hence the issue of animal rights does not arise.

Besides the target protein, the by-products that can be obtained are rubber and, eventually, cultivated tropical timber, both of which are valuable commodities in their own rights.

We thank the Director, Rubber Research Institute of Malaysia, for permission to publish this paper. Agrobacterium tumefaciens GV2260 [p35SGUSINT] was a gift from Professor L. Willmitzer, Institut für Genbiologische Forschung, Berlin. Dr. S.K. Leong facilitated the vegetative multiplication of the transgenic plants by bud-grafting. Technical assistance in the transformation experiments from S. Rajamanikam, N.P. Chew and J. Sharib is gratefully acknowledged. The experimental work was jointly funded by the Ministry of Science, Technology and the Environment, Malaysia, through IRPA grant 1811002004-0004 and by the British Council, Malaysia through the British High Commissioner's Award to P.A.

References

1. Gomez, J.B. (1982) Anatomy of *Hevea* and its Influence on Latex Production, Malaysian Rubber Research and Development Board, Kuala Lumpur
2. Buttery, B.R. and Boatman, S.G. (1964) Science **145**, 285–286
3. Buttery, B.R. and Boatman, S.G. (1966) J. Exp. Bot. **17**, 283–296
4. Boatman (1966) J. Rubb. Res. Inst. Malaya **19**, 243–258
5. Statistics Department of Malaysia (1988) Rubber Statistical Handbook, Statistics Department of Malaysia, Malaysia
6. Statistics Department of Malaysia (1995) Rubber Statistical Handbook, Statistics Department of Malaysia, Malaysia
7. Statistics Department of Malaysia (1971) Rubber Statistical Handbook, Statistics Department of Malaysia, Malaysia
8. Tong, K.H. (1995) in Ensuring Sustainability and Competitiveness of the NR Industry (Abdul Aziz, S.A.K., ed.), pp. 287–308, Rubber Research Institute of Malaysia, Kuala Lumpur
9. Archer, B.L., Barnard, D., Cockbain, E.G., Dickenson, P.B. and McMullen, A.I. (1963) in The Chemistry and Physics of Rubber-like Substances (Bateman, L., ed.), pp. 43–72, Maclaren & Sons Ltd., London, and Wiley, New York
10. Archer, B.L., Audley, B.G., McSweeney, G.P. and Tan, C.H. (1969) J. Rubb. Res. Inst. Malaya **21**(4), 560–569

11. Kageyama, K. (1993) in Natural Rubber: Current Developments in Product Manufacture and Applications (Abdul Aziz, S.A.K., ed.) pp. 84–97, Rubber Research Institute of Malaysia, Kuala Lumpur
12. Yeang, H.Y., Arokiaraj, P., Cheong, K.F., Coomber, S. and Charlwood, B.V. (1995) Abstracts AAAS Ann. Gen. Meeting and Sci. Innovation Expo. 16–21 February 1995, Atlanta, p. 153
13. Begley, S. (1991) Newsweek, 9 September, p. 50.
14. Gomez, J.B. and Moir, G.F.J. (1979) in The Ultracytology of Latex Vessels in *Hevea brasiliensis*, pp. 39–40, Malaysian Rubber Research and Development Board, Kuala Lumpur
15. Dickenson, P.B. (1965) in Proc. Nat. Rubb. Prod. Res. Assoc. Jubilee Conf., Cambridge, 1964 (Mullins, L. ed.), pp. 52–66, Maclaren & Sons Ltd, London
16. Brem, G., Hartl, P., Besengelder, U., Wolf, E., Zinovieva, N. and Pfaller, R. (1994) Gene **149**, 351–355
17. Wilmut, T.I. and Whitelaw, C.B.A. (1994) Reprod. Fertil. Dev. **6**, 625–630
18. Ward, K.A. and Nancarrow, C.D. (1995) Mol. Biotechnol. **4**, 167–178
19. Houdebine, L.M. (1995) Reprod. Nutr. Dev. **35**, 609–617
20. Coleman, A. (1996) Am. J. Clin. Nutr. **63**, 639S–645S
21. Ong, T.S., Heh, W.Y. and Wong, C.P. (1991) Proc. Rubb. Res. Inst. Malaysia Rubb. Grow. Conf. Malacca 1989, pp. 110–124, Rubber Research Institute of Malaysia, Kuala Lumpur
22. Moir, G.F.J. (1959) Nature (London) **21**, 1626–1628
23. d'Auzac, J. and Jacob, J.L. (1969) J. Rubb. Res. Inst. Malaya, **21**(4), 417–444
24. Bealing, F.J. (1969) J. Rubb. Res. Inst. Malaya, **21**(4), 445–455
25. Yeang, H.Y., Yip, E. and Samsidar Hamzah (1995) J. Nat. Rubb. Res. **10**(2), 108–123
26. Tata, S.J. (1980) Studies on the Lysozyme and Components of Microhelices of *Hevea brasiliensis* Latex, PhD Thesis, University of Malaya, p. 75
27. Abraham, P.D., Wycherley, P.R. and Pakianathan, S.W. (1968) J. Rubb. Res. Inst. Malaya **20**, 291–305
28. Abraham, P.D., Blencowe, J.W., Chua, S.E., Gomez, J.B., Moir, G.F.J., Pakianathan, S.W., Sekhar, B.C., Southorn, W.A. and Wycherley, P.R. (1971) J. Rubb. Res. Inst. Malaya **23**, 90–113
29. Poirier, Y.P., Dennis, D., Klomparens, K., Nawrath, C. and Somerville, C. (1992) FEMS Microbiol. Rev. **103**, 237–246
30. Yeang, H.Y. and Chevallier, M.H. (1992) in Abstracts Fourth National Biotechnology Seminar, Selangor, Malaysia, pp. 73–74
31. Arokiaraj, P. and Wan Abdul Rahaman, W.Y. (1991) J. Nat. Rubb. Res. **6**, 55–61
32. Kitayama, M., Takahashi, M., Surzycki, S.J. and Togasaki, R.K. (1990) Plant Physiol. **93** (Suppl.), 46
33. Jefferson, R.A. (1987) Plant Mol. Biol. Rep., **5**, 387–405
34. Arokiaraj, P., Jones, H., Cheong, K.F., Coomber, S. and Charlwood, B.V. (1994) Plant Cell Rep. **13**, 425–431
35. Vancanneyt, G., Schmidt, R., O'Connor-Sanchez, A., Willmitzer, L. and Roch-Sosa, M. (1990) Mol. Gen. Genet., **220**, 245–250
36. Arokiaraj, P., Yeang, H.Y., Cheong, K.F., Hamzah, S., Jones, H., Coomber, S. and Charlwood, B.V. (1998) Plant Cell Rep. **17**, 621–625
37. Z. Chen (1984) in Handbook of Plant Cell Culture. Crop Species, Vol. 2: Rubber (Sharp, W.A., Evans, D.A., Ammirato, P.V. and Yamada, Y., eds.), pp. 546–571, Macmillan, New York
38. Jefferson, R.A., Kavanagh, T.A. and Bevan, M.W. (1987) EMBO J. **6**, 3901–3907

Food protein engineering of soybean proteins and the development of soy-rice

Tomoyuki Katsube*§, Nobuyuki Maruyama*, Fumio Takaiwa† and Shigeru Utsumi*‡
*Research Institute for Food Science, Kyoto University, Uji, Kyoto 611, Japan and
†National Institute of Agrobiological Resources, Tsukuba, Ibaraki 305, Japan

Introduction

Soybean (*Glycine max* L.) has a long history as a protein foodstuff in the orient, including Japan. Over the centuries a variety of soybean foods such as tofu, abura-age, koori-dofu and yuba have been developed. In Japan, the consumption of soybean is around 5 million tonnes, with 80% used for oil expression, 11% for tofu and abura-age, and 1% for koori-dofu. Besides such traditional utilization, a number of soybean products have been developed in recent times, including Soy Kara-Aghe (looking and tasting like chicken nugget), which is made by extrusion of soybean protein isolate and frying. This commercial product is approved as 'Food for specified health use' in Japan because of its ability to lower cholesterol levels in human serum.

The annual production of soybean seeds in the world is 120 million tonnes with more than 80% being used for oil expression. The amount of the protein in the residues after oil expression reaches around 35 million tonnes. Some 1.7 billion people can survive for one year on this amount of protein, assuming that it is consumed directly and the utilization efficiency is 60%. However, most of the residues are used as feed for domestic animals or discarded, since the proteins present do not have the functional properties required for the production of attractive processed foods. In addition, soybean proteins are deficient in the essential sulphur-containing amino acids. Therefore, improvement of the nutritional value and functional properties (gel-forming and emulsifying abilities are significant functional properties with respect to utilization in food systems) of soybean proteins is a major objective of food scientists and their co-workers in helping to solve the problem of food shortages in the near future.

Soybean seeds contain glycinin (11 S globulin) and β-conglycinin (7 S globulin) as major components, accounting for about 40 and 30% of the total proteins, respectively [1]. Glycinin is responsible for the cholesterol-lowering properties of soybean proteins and is generally superior to β-conglycinin with

‡To whom correspondence should be addressed.
§Shimane Women's College, Matsue, Shimane 690, Japan.

respect to nutritional and functional properties [1]. Therefore, glycinin is a suitable target for the creation of an ideal food protein, while β-conglycinin also needs to be modified to improve its functional properties.

Food protein engineering of glycinin

Glycinin is a simple protein with a hexameric structure. The constituent subunits of glycinin are synthesized as a single polypeptide precursor (preproglycinin) [2–4]. The signal sequence is processed co-translationally in the endoplasmic reticulum (ER), and the resultant proglycinin assemble into trimers of about 8 S [3–5]. These complexes are sorted from the ER to the vacuoles, where a specific post-translational cleavage occurs [6]. The cleavage results in mature subunits, each consisting of an acidic and a basic polypeptide linked by a disulphide bridge [7], that assemble into hexamers of about 12 S [3–5]. Finally, glycinins accumulate in a densely packed state in protein bodies. Molecular assembly, sorting into the vacuoles and accumulation in protein bodies are defined by topogenic information contained in the glycinin molecule. Accordingly, the modifications introduced into a glycinin molecule by protein engineering to improve food functionality should not result in malfolding when expressed in novel crops [1]. Since it takes a long time to produce and examine many transgenic plants, it is desirable to evaluate whether modified glycinins are able to adopt a conformation similar to that of the native protein and to exhibit the expected functional properties before the genes are transferred into crops. The establishment of a high-level expression system for glycinin cDNA in a micro-organism is therefore desirable.

Construction of expression systems for glycinin cDNA in micro-organisms

Five types of subunit have been identified as constituents of glycinin and are classified into two groups: group I (A1aB1b, A1bB2, A2B1a) and group II (A3B4, A5A4B3) [1]. The group I subunits have better nutritional value than those of group II. Therefore, a subunit belonging to group I is a suitable target for improvement to create an ideal food protein.

Expression of cDNA encoding the glycinin A1aB1b subunit in *Escherichia coli* was attempted using the expression vector pKK233-2. We could not detect glycinin proteins from cDNA encoding the prepro-form of the subunit [8]. This phenomenon seems to be due to disturbance of the folding of the expressed proteins by the presence of the hydrophobic signal sequence that is not cleaved in *E. coli* and the resulting susceptibility of the protein to proteinase digestion. This was confirmed by stepwise deletion of the cDNA sequence encoding the signal sequence and the mature N-terminal region of the A1aB1b cDNA: no accumulation of expressed protein was observed from the plasmids for proteins with either the full-length signal sequence or with five amino acids of the signal sequence while significant accumulation was observed for proteins retaining less than three amino acids of the signal sequence [9]. The highest expression level (20% of the total *E. coli* proteins) was obtained using the expression plasmid for proglycinin A1aB1b-3 lacking three amino acids from the mature N-terminus with optimized culture conditions (37 °C, shaken at 90 strokes·min^{-1}) [10]. *E. coli* cells do not contain an enzyme able to cleave proglycinin to a mature form [9]. Consequently, the proteins expressed from the proglycinin cDNA accumulate as proglycinin in *E. coli*. The proglycinin A1aB1b-3 subunits expressed at 37 °C are soluble, self-assemble into trimers and have a similar secondary structure to that of mature glycinin [10], suggesting that the proglycinin A1aB1b-3 does not require ER lumenal proteins such as BiP (a molecular chaperone) and protein disulphide isomerase for correct folding. The expressed proglycinin A1aB1b-3 exhibits similar fundamental properties to glycinin; such as cryoprecipitation, calcium-induced precipitation, heat-induced gelation and emulsification [10]. Therefore, the *E. coli* expression system employing pKK233-2 and JM105 can be used to evaluate the formation of the correct conformation and to determine the functional properties of engineered glycinin, although the expressed protein is the pro-form not the mature form.

Construction of a yeast (*Saccharomyces cerevisiae*) expression system was attempted using an expression vector (pAM82) containing the repressible acid phosphatase promoter *PHO5* [11,12]. The signal sequence of the expressed protein was correctly processed at the same site as in soybean and the expression level reached around 5% of the total yeast proteins. However, since yeast does not contain a maturation protease enzyme, the expressed protein is again the pro-form. Most of the expressed protein was insoluble, owing to interaction between the acidic polypeptide region and other intracellular components [12]. Although renaturation from the reduced-denatured state to trimer structure is possible after partial purification [12], the *E. coli* expression system is considered to be superior to the yeast system for routine protein engineering of glycinin.

Strategy for design of modified glycinins with enhanced food functionality

To improve the nutritional and functional properties of glycinin by protein engineering, the following two points should be considered: (1) which regions of the glycinin molecule can tolerate modifications by protein engineering? and (2) what kinds of modification are effective to improve the properties?

The three-dimensional structure determined by X-ray crystallography is an ideal starting point. However, when we started these studies such data were not available. Wright [13] aligned the amino acid sequences to maximize the homology among the 11 S globulins from various legumes and non-legumes, and proposed that they comprised a series of alternating conserved and variable regions. The variable regions are generally hydrophilic, suggesting that they are located on the surface of the protein. It is probable that the variable regions have little function in forming and maintaining the glycinin structure and may therefore tolerate modification. Glycinin contains five variable regions (Figure 1).

The strategy for improvement of the functional properties can be based on our knowledge of the relationship between the structure and the functional properties of glycinin. Nakamura and co-workers [14] proposed, based on a study of the relationship between the structure at the subunit level and heat-induced gelation of glycinin, that the thermal instability of the constituent subunits was

Figure 1

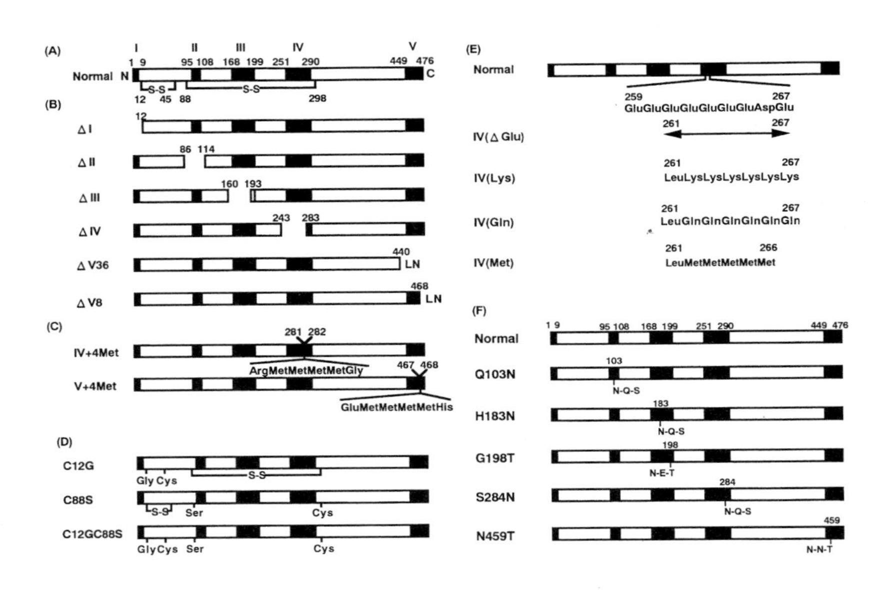

Schematic representation of the normal and modified proglycinins

Black and open areas are variable and conserved regions, respectively. N and C represent the N- and C-terminus, respectively. The numbers of residues from the N-terminus for the variable regions I–V are shown above the alignment. A, Schematic representation of the normal proglycinin A1aB1b. The positions of the disulphide bonds are indicated. B, Deletion mutants lacking individual variable regions. ΔV36 and ΔV8 have two extra amino acids, Leu-Asn, derived from the universal terminator at their C-terminus. C, Insertion mutants having tetramethionines. D, Disulphide bond-deleted mutants. E, Deletion and substitution mutants having higher isoelectric points than that of the normal proglycinin A1aB1b. F, Mutants having an N-glycosylation consensus sequence. Figure created from [18] with permisson from Oxford University Press, from [23] with permission from the American Chemical Society (©1994), from [1] with permission and from S. Utsumi and M. Kito (1991) Comments Agric. Food Chem. **2**, *261–278, with permission.*

related to the heat-induced gel-forming ability: destabilization and stabilization should therefore promote and inhibit gelation, respectively. The number and the topology of free sulphydryl groups as well as disulphide exchange reaction are closely related to the heat-induced gel-forming ability and the gel properties of glycinin [1,14,15]. On the other hand, attachment of fatty acids to glycinin increased its emulsifying properties [16]. It was reported that the surface properties of a protein depend on the conformational stability: the more unstable the conformation, the higher the emulsifying properties [17]. These facts suggest that (1) deletion or substitution for other sequences of the variable regions, (2) insertion of multiple hydrophobic amino acids into the variable regions, (3) the substitution of a cysteine residue involved in a disulphide bond with another amino acid and (4) glycosylation, may be powerful approaches to improve or change the functional properties of glycinin. This is because such modifications may result in changes in the hydrophobicity or the hydrophilicity, in alterations in the number and the topology of free sulphydryl groups and, consequently, in the destabilization or stabilization of the glycinin molecule. Improvement of the nutritional value of glycinin can be achieved by enrichment with the limiting amino acid methionine. Based on these ideas, we designed 20 modified glycinins, as shown in Figure 1. The strategies used are classified into five groups.

Deletion of individual variable regions

Since the variable regions are strongly hydrophilic, removal of individual variable regions results in a strengthening of the relative hydrophobicity. The removal of these regions from the glycinin molecule would result in its destabilization. Consequently, we could expect improvements in the heat-induced gel-forming and emulsifying abilities. Therefore, deletion mutants lacking each individual variable region were designed as shown in Figure 1(B): ΔI lacks the N-terminal eleven amino acids, ΔII from residue 87 to 113, ΔIII from 161 to 192, ΔIV from 244 to 282, ΔV36 from 441 to the C-terminus and ΔV8 from 469 to the C-terminus [18].

Insertion of oligopeptides into variable regions

Methionine, the nutritionally limiting amino acid of soybean, has a hydrophobic nature. Insertion of an oligopeptide composed of contiguous multiple methionines into the variable regions results in enhancement of hydrophobicity and destabilization of the glycinin molecule. Consequently, improvement of the gel-forming and emulsifying abilities can also be expected in addition to enrichment of the nutritional value. Thus, insertion mutants IV+4Met and V+4Met were designed (Figure 1C) [18]. Arg-Met-Met-Met-Met-Gly and Glu-Met-Met-Met-Met-His were inserted between Pro-281 and Arg-282 and between Pro-467 and Glu-468 in the fourth and the fifth variable regions, respectively. The insertions result in increased hydrophobicity of the insertion sites [19].

Deletion of disulphide bond(s)

Disulphide bonds contribute to the maintenance of protein structure. Deletion of disulphide bond(s) by the substitution of a cysteine residue involved in disulphide bond formation with another amino acid alters the number and topology of the free sulphydryl groups and also destabilizes the glycinin molecule. Therefore, the gel-forming and emulsifying abilities should be improved. Individual constituent subunits of glycinin have two or three disulphide bonds [20]. Two disulphide bonds Cys-12-Cys-45 and Cys-88-Cys-298 have been identified in the A1aB1b subunit [1,7]. Accordingly, the disulphide bond deletion mutants C12G and C88S were designed (Figure 1D) [21]. Cys-12 was substituted with Gly in C12G and Cys-88 with Ser in C88S. As a result of the substitutions, C12G and C88S have new free sulphydryl groups at positions 45 and 298, respectively. C12GC88S with two new free cysteine residues at positions 45 and 298 was also designed.

Deletion and substitution of the polyglutamic acid sequence

The sequence between residues 259 and 267 in the fourth variable region is composed of one aspartic acid and eight glutamic acid residues (polyglutamic acid sequence). By altering the composition of the polyglutamic acid sequence, the isoelectric point of the subunit can be changed. Soybean proteins do not show their maximum functional properties in the acidic range (pH 3–6), because their solubility decreases owing to their low isoelectric points [22]. Therefore, alteration of the net electric charge of the glycinin molecule would allow for the utilization of soybean proteins for acidic foods such as mayonnaise and yogurt. Thus, deletion mutant IV(ΔGlu) and substitution mutants IV(Lys), IV(Gln) and IV(Met) were designed (Figure 1E) [23]. The sequence between positions 261 and 266 or 267 are deleted for IV(ΔGlu) and substituted with Leu-Lys-Lys-Lys-Lys-Lys-Lys, Leu-Gln-Gln-Gln-Gln-Gln-Gln and Leu-Met-Met-Met-Met-Met for IV(Lys), IV(Gln) and IV(Met), respectively. The hydropathy profile changes significantly as a result of the deletion of the polyglutamic acid sequence and the substitution with a polymethionine sequence [23].

Introduction of an *N*-glycosylation consensus sequence

It has been suggested that glycosylation is one of the most promising approaches to increase stability to heating [24]. Thus, an N-glycosylation consensus sequence Asn-Xaa-Ser/Thr was created at positions 103, 183, 196, 284 and 457 in each variable region to construct Q103N, H183N, G198T, S284N and N459T (Figure 1F) [24a].

Evaluation of food functionality of modified proglycinins

The modified proglycinins should be able to form the correct conformation, similar to that of native proglycinin. To confirm that the conformation was correct we employed the following three criteria: (1) the solubility should be comparable with that of the native protein, (2) there must be self-assembly into trimers and (3) the protein should be stable under conditions of high ionic strength [18,21,25]. These criteria were applied to all the modified proglycinins (shown in Figures 1B–1E) expressed in *E. coli* cells. Among the modified proglycinins expressed in *E. coli*, ΔI, ΔV8, IV+4Met, V+4Met, C12G, C88S, C12GC88S, IV(ΔGlu), IV(Lys), IV(Gln) and IV(Met) were all shown to have similar conformations to that of the native proglycinin [18,21,23]. These 11 modified proglycinins were purified by ammonium sulphate fractionation and Q-Sepharose column chromatography. All the modified proglycinins were purified to near homogeneity except C12GC88S, which is susceptible to attack by proteinases at low ionic strength [21]. All the purified normal and modified proglycinins expressed in *E. coli* can form crystals under suitable conditions [23,26,27]. This confirms that these modified proglycinins can form a conformation similar to that of the native proglycinin, and indicates that each modified proglycinin has different structural characteristics that may affect its functional properties. On the other hand, yeast expression plasmids were constructed for the modified proglycinins having the N-glycosylation consensus sequence shown in Figure 1(F). Among the five modified proglycinins shown in Figure 1(F), only Q103N was fully glycosylated, the others being not glycosylated or only partly glycosylated [24a]. Q103N was partly purified by Q-Sepharose column chromatography under reducing and denaturing conditions, and then renatured by two-step dialysis. The renatured Q103N assembles into trimers [24a]. The Q103N trimers were purified to near homogeneity by Q-Sepharose column chromatography.

The functional properties of the purified modified proglycinins were determined. The isoelectric points of IV(ΔGlu), IV(Lys), IV(Gln) and IV(Met) were measured by isoelectric focusing, in relation to their solubility at acidic pH. The values were 6.6, 7.2, 6.5 and 6.4, respectively, which are significantly higher than that (5.6) of the normal proglycinin [23]. Therefore, we can expect that these modified proglycinins, especially IV(Lys), should show good functional properties in the acidic pH range. The emulsifying and heat-induced gel-forming properties of ΔI, ΔV8, IV+4Met, V+4Met, C12G and C88S were compared with those of native glycinin from soybean and the normal proglycinin [18,21]. All the modified proglycinins exhibited higher emulsifying activities than the native soybean glycinin, especially ΔV8 and V+4Met, which were twice as active as native glycinin. This result suggests that the hydrophobicity of the C-terminal region may be closely related to the emulsifying properties of glycinin. Meanwhile, all the modified proglycinins formed gels on boiling. The unmodified proglycinin formed gels that were slightly softer than those formed by native glycinin. The gels from C12G were of similar hardness to the native glycinin gels at higher protein concentrations (>6 %), but C12G could not form gels at lower protein concentrations (<5.6 %). This suggests that the disulphide bond C12–C45 plays an important role in the initiation of SH/S–S exchange reactions in gelation. The ΔV8 gels were less hard. Nevertheless, the gels from ΔI, IV+4Met, V+4Met and C88S were harder than the native glycinin gels; notably, C88S formed hard gels even at a low protein concentration (4.4%), where the native glycinin formed very soft gels. In addition, the thermal stability of Q103N was compared with that of the normal proglycinin. The result indicates that Q103N is more thermostable than the normal proglycinin. In other words, glycosylated glycinin is suitable for liquid food requiring sterilization [24a].

Evaluation of the functional properties of the modified proglycinins demonstrated that most had better properties for at least one food use than the native glycinin. It is especially noteworthy that IV+4Met and V+4Met exhibited better emulsifying and gel-forming abilities as well as higher nutritional value. Thus, IV+4Met and V+4Met are greatly improved food proteins.

Food protein engineering of β-conglycinin

β-Conglycinin is a glycoprotein having a trimeric structure, being composed of three types of subunit, α, α′ and β. Each subunit is synthesized as a preproprotein (α and α′) or a preprotein (β). The primary structures of these subunits resemble each other, especially in the core regions (Figure 2). The sequence identities in the core regions between α and α′, α and β, and α′ and β are 86.8, 75.5 and 71.4%, respectively [28–30]. Extension regions of the α and α′ subunits exhibit 57.3% sequence identity and contain high proportions of acidic amino acids. Therefore, deletion of the extensions gives core regions (αc and α′c) of α and α′ that exhibit dramatically different properties to the whole subunits. The molecular masses, isoelectric points, hydrophobicities [31] and contents of essential amino acids (methionine, cysteine and tryptophan) calculated from the amino acid compositions of the normal subunits and deletion mutants (αc and α′c) are

Figure 2

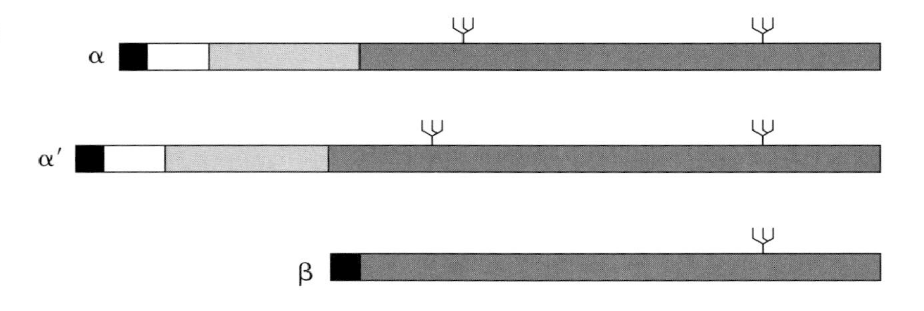

Schematic representation of constituent subunits of β-conglycinin

Black, open, hatched and dotted areas are signal-sequence, pro-sequence, extension region and core region, respectively. N-linked glycans are depicted by the branched structure.

summarized in Table 1. The deletion mutants αc and α′c have different isoelectric points and hydrophobicities to α and α′ and contain no cysteine residues. Although the hydrophobicities of αc and α′c are similar to that of β, their isoelectric points are much higher. Therefore, we can expect that αc and α′c will exhibit different functional properties to those of α, α′ and β. Moreover, the contents of methionine in αc and α′c are the same as in the whole subunits, although αc and α′c contain no tryptophan residues. From this viewpoint, we are currently trying to compare the functional properties of α, α′, β, αc and α′c expressed in *E. coli*.

Development of soy-rice

The major storage protein of rice is glutelin, which accounts for 80% of the rice storage proteins. Each constituent subunit of glutelin is synthesized as a single polypeptide and processed to a mature form consisting of an acidic and a basic polypeptide, similar to the constituent subunits of glycinin and other glycinin-type proteins (pea legumin, field bean legumin, oat 12 S globulin, etc.) [13]. The constituent subunits of glutelin are classified into two groups (glutelin A and B) on the basis of their amino acid sequences. Generally, glutelins A and B contain eight and five cysteine residues, respectively. Four of these cysteine residues correspond to the four cysteine residues of the glycinin subunits that form intra-subunit disulphide bonds (Figure 3) [7]. These four cysteine residues are conserved in equivalent positions in all glycinin-type proteins so far sequenced

Table 1

	Molecular mass (kDa)	pI	Hydrophobicity	Essential amino acids Trp	Cys	Met
α	63	4.73	−1.06	1	1	1
αc	48	9.15	−0.54	0	0	1
α′	67	5.07	−1.13	2	1	4
α′c	50	7.72	−0.56	0	0	4
β	48	5.48	−0.61	0	0	0

Characteristics of β-conglycinin subunits and mutants

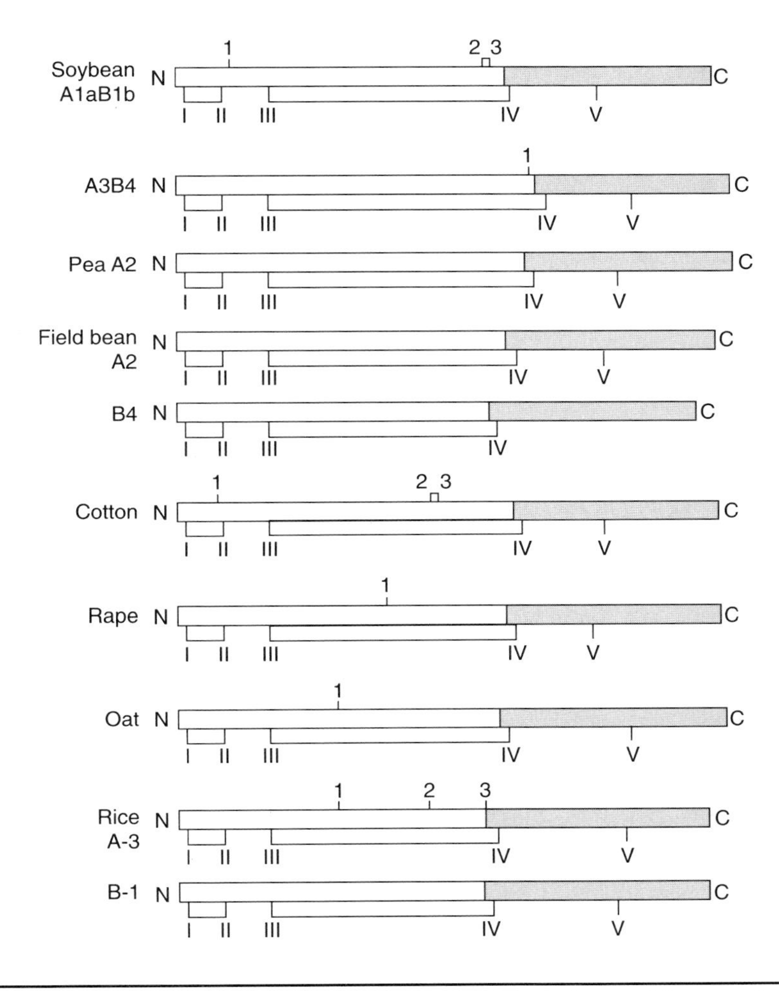

Schematic representation of the proposed positions of free cysteine residues and disulphide bonds in glycinin, glutelin and glycinin-type proteins

Open and hatched areas are the acidic and the basic polypeptides, respectively. N and C represent the N- and C-terminus, respectively. The cysteine residues conserved and not-conserved among glycinin-type proteins are numbered by Roman and Arabic numerals, respectively. Sequence data are from EMBL accession numbers M36686, M10962, X17193, X55014, X03677, M69188, J05233, X17637, X54313 and X54314.

[20]. These facts suggest that glutelins A and B also have intra-subunit disulphide bonds in the same positions as the other glycinin-type proteins. The position of the cysteine residue V of glutelin B corresponds to that of the free cysteine residue V of other glycinin-type proteins. The oat 12 S globulin also has cysteine residue V in the same position as in the glutelin, but this residue is absent from the field bean legumin B4. This position is also variable among other glycinin-type proteins (Figure 3). These suggest that cysteine residue V of glutelin B may be present as a free sulphydryl group. Therefore, it is likely that glutelin A plays an

Figure 4

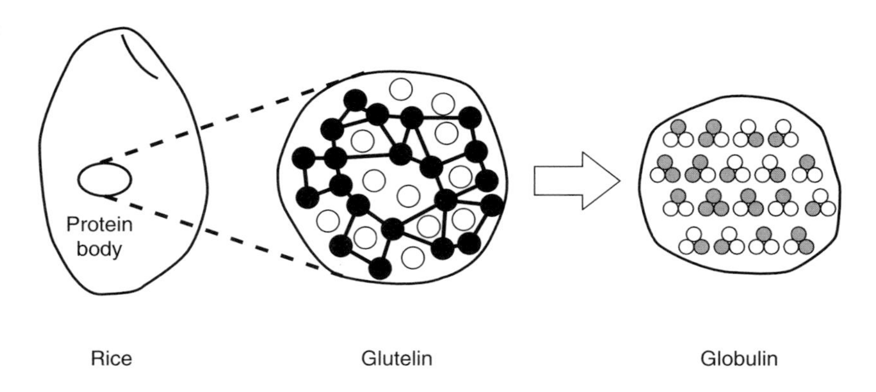

Conceptual diagram of protein association in rice protein bodies and conversion of the rice storage proteins glutelin into globulin

Black, open and shaded circles represent glutelins A, B and glycinin subunits, respectively. The lines connecting black circles represent disulphide bonds.

important role in the formation of macromolecules consisting of disulphide-bonded subunits, as shown in Figure 4. If so, it is possible that glutelin B would be able to assemble into hexamers together with glycinin subunits if the modified glycinin gene was transformed into rice and the expression of glutelin A genes suppressed, resulting in conversion of the properties of rice storage proteins from glutelin into globulin.

Although glutelin subunits and glycinin subunits have a common fundamental structure, glutelin exhibits no significant functional properties. Glutelin is deficient in lysine whereas glycinin is rich in this amino acid. Therefore, conversion of rice storage proteins from glutelin into globulin will furnish them with functional properties and improve their nutritional quality. The development of an alternative to rice will also help to eliminate the problem of off-flavours in rice stored for long periods, which is a serious problem in Japan.

To facilitate the creation of soy-rice, we initially examined the expression of normal and modified glycinin genes under the control of glutelin promoter in tobacco seeds. The normal and modified glycinin cDNAs were fused to the 5′ flanking region (1320 bp) of the glutelin *Glu B-1* gene [32] and then inserted into the binary vector pBI101. The chimaeric constructs were introduced into the tobacco genome by *Agrobacterium*-mediated transformation [33]. Twenty-three and 34 independent plants were regenerated for the normal form and the V+4Met mutant, respectively, and the expression levels determined immunochemically. In both cases, more than 60% of plants accumulated the normal and modified glycinins at a level of >1% of the total proteins in the dry seeds. The highest level was about 4% for both the normal and V+4Met. This level was more than 40-fold higher when compared with that directed by the CaMV35S promoter [34], and compares well than those reported for other systems (0.003–2% of total seed proteins [35–40]). The normal and modified glycinins were processed into mature-forms, assembled into hexamers and accumulated in the protein bodies in the endosperm tissue, although about half of the synthesized glycinins were partially digested and processing and assembly were insufficient [33]. Similar degradation is often observed when heterologous storage proteins are expressed in transgenic plants, regardless of the promoter [33]. These results indicate that the combination of glutelin promoter and glycinin coding region is suitable for our purpose and that it is possible to develop soy-rice producing modified glycinins with enhanced food quality (nutritional and functional properties). We are now trying to create soy-rice, and are obtaining good results.

This work was supported in part by Grants from the Ministry of Education, Science and Culture of Japan (S.U. and T.K.), from the Ministry of Agriculture, Forestry and Fisheries of Japan (S.U. and F.T.), Asahi Bear Foundation (S.U.), The Iijima Memorial Foundation for the Promotion of Food Science and Technology (S.U.), The Skylark Food Science Institute (S.U.), The Soy Protein Research Committee (S.U.), Takano Life Science Research Foundation (S.U.) and PROBRAIN (S.U. and F.T.).

References

1. Utsumi, S. (1992) in Advances in Food and Nutrition Research, Vol. 36 (Kinsella, J.E., ed.), pp. 89–208, Academic Press, San Diego
2. Staswick, P.E., Hermodson, M.A. and Nielsen, N.C. (1981) J. Biol. Chem. **256**, 8752–8755
3. Barton, K.A., Thompson, J.F., Madison, J.T., Rosenthal, R., Jarvis, N.P. and Beachy, R.N. (1982) J. Biol. Chem. **257**, 6089–6095
4. Tumer, N.E., Richter, J.D. and Nielsen, N.C. (1982) J. Biol. Chem. **257**, 4016–4018
5. Chrispeels, M.J., Higgins, T.J.V. and Spencer, D. (1982) J. Cell Biol. **93**, 306–313
6. Nielsen, N.C. (1984) Philos. Trans. R. Soc. London, Ser. B **304**, 287–296
7. Staswick, P.E., Hermodson, M.A. and Nielsen, N.C. (1984) J. Biol. Chem. **259**, 13431–13435
8. Utsumi, S., Kim, C-S., Kohno, M. and Kito, M. (1987) Agric. Biol. Chem. **51**, 3267–3273
9. Utsumi, S., Kim, C-S., Sato, T. and Kito, M. (1988) Gene **71**, 349–358
10. Kim, C-S., Kamiya, S., Kanamori, J., Utsumi, S. and Kito, M. (1990) Agric. Biol. Chem. **54**, 1543–1550
11. Utsumi, S., Sato, T., Kim, C-S. and Kito, M. (1988) FEBS Lett. **233**, 273–276
12. Utsumi, S., Kanamori, J., Kim, C-S., Sato, T. and Kito, M. (1991) J. Agric. Food Chem. **39**, 1179–1186
13. Wright, D.J. (1988) in Developments in Food Proteins (Hudson, B.J.F., ed.), pp. 119–178, Elsevier, London
14. Nakamura, T., Utsumi, S., Kitamura, K., Harada, K. and Mori, T. (1984) J. Agric. Food Chem. **32**, 647–651
15. Mori, T., Nakamura, T. and Utsumi, S. (1982) J. Food Sci. **47**, 26–30
16. Utsumi, S. and Kito, M. (1991) Comments Agric. Food Chem. **2**, 261–278
17. Kato, A. and Yutani, K. (1988) Protein Eng. **2**, 153–156
18. Kim, C-S., Kamiya, S., Sato, T., Utsumi, S. and Kito, M. (1990) Protein Eng. **3**, 725–731
19. Utsumi, S., Katsube, T., Ishige, T. and Takaiwa, F. (1997) in Food Proteins and Lipids (Damodaran, S., ed.), pp. 1–15, Plenum Press, New York
20. Utsumi, S., Matsumura, Y. and Mori, T. (1997) in Food Proteins and Their Applications (Damodaran, S. and Paraf, A., eds.), pp. 257–291, Marcel Dekker, New York
21. Utsumi, S., Gidamis, A.B., Kanamori, J., Kang, I.J. and Kito, M. (1993) J. Agric. Food Chem. **41**, 687–691
22. Kinsella, J.E. (1979) J. Am. Oil Chem. Soc. **56**, 242–258
23. Katsube, T., Gidamis, A.B., Kanamori, J., Kang, I.J., Utsumi, S. and Kito, M. (1994) J. Agric. Food Chem. **42**, 2639–2645
24. Kato, A., Takasaki, H. and Ban, M. (1994) FEBS Lett. **355**, 76–80

24a. Katsube, T., Kang, I.J., Takenaka, Y., Adachi, M., Maruyama, N., Morisaki, T. and Utsumi, S. (1998) Biochim. Biophys. Acta **1379**, 107–117

25. Gidamis, A.B., Wright, P., Haque, Z.U., Katsube, T., Kito, M. and Utsumi, S. (1995) Biosci. Biotechnol. Biochem. **59**, 1593–1595
26. Utsumi, S., Gidamis, A.B., Mikami, B. and Kito, M. (1993) J. Mol. Biol. **233**, 177–178

27. Gidamis, A.B., Mikami, B., Katsube, T., Utsumi, S. and Kito, M. (1994) Biosci. Biotechnol. Biochem. **58**, 703–706
28. Doyle, J.J., Schuler, M.A., Godette, W.D., Zenger, V., Beachy, R.N. and Slightom, J.L. (1986) J. Biol. Chem. **261**, 9228–9238
29. Harada, J.J., Barker, S.J. and Goldberg, R.B. (1989) Plant Cell **1**, 415–425
30. Sabastiani, F.L., Farrell, L.B., Schuler, M.A. and Beachy, R.N. (1990) Plant Mol. Biol. **15**, 197–201
31. Kyte, J. and Doolittle, R.F. (1982) J. Mol. Biol. **157**, 105–132
32. Takaiwa, F., Oono, K., Wing, D. and Kato, A. (1991) Plant Mol. Biol. **17**, 875–885
33. Takaiwa, F., Katsube, T., Kitagawa, S., Higasa, T., Kito, M. and Utsumi, S. (1995) Plant Sci. **111**, 39–49
34. Utsumi, S., Kitagawa, S., Katsube, T., Kang, I.J., Gidamis, A.B., Takaiwa, F. and Kito, M. (1993) Plant Sci. **92**, 191–202
35. Beachy, R.N., Chen, Z.L., Horsch, R.B., Rogers, S.G., Hoffmann, N.J. and Fraley, R.T. (1985) EMBO J. **4**, 3047–3053
36. Sengupta-Gopalan, C., Reichert, N.A., Barker, R.F., Hall, T.C. and Kemp, J.D. (1985) Proc. Natl. Acad. Sci. U.S.A. **82**, 3320–3324
37. Higgins, T.J.V., Newbigin, E.J., Spencer, D., Llewellyn, D.J. and Craig, S. (1988) Plant Mol. Biol. **11**, 683–695
38. Williamson, J.D., Galili, G., Larkins, B.A. and Gelvin, S.B.G. (1988) Plant Physiol. **88**, 1002–1007
39. Robert, L.S., Thompson, R.D. and Flavell, R.B. (1989) Plant Cell **1**, 569–578
40. Bogue, M.S., von der Haar, R.A., Nuccio, M.L., Griffing, L.R. and Thomas, T.L. (1990) Mol. Gen. Genet. **221**, 49–57

Developing novel pea starches

Cliff L. Hedley*, Tanya Ya. Bogracheva and Trevor L. Wang
John Innes Centre, Norwich Research Park, Norwich NR4 7UH, U.K.

Why peas and why pea starches?

Potential for the pea crop

Over the past 20 years there has been a rapid increase in the production of peas for the dried pea seed market within the E.U. and, in 1994, this production reached about 5 million tonnes. The increase was partly due to the application of subsidies, imposed by the Commission to boost the production of home-grown, high-protein crops, and partly to the development of new varieties. The major problem with the pea crop has always been lodging late in the season, making the crop prone to disease and difficult to harvest. New so-called 'leafless' varieties, incorporating the *afila* gene, improved the standing ability of the crop, mainly by the increased number of leaf tendrils interlinking neighbouring plants to provide mutual support. These new varieties have made the pea crop more reliable with regard to harvesting and less variable in terms of yield. In France the average yield of dried peas over the past two years has been in excess of 5 tonnes per hectare and now crops of more than 7 tonnes per hectare are not uncommon. About 90% of dried peas within the E.U. is used by processors to produce high-protein animal feedstuff, less than 4% is used for human food and an even smaller proportion as a source of raw materials [1]. Now that the crop has become more established with growers there is a great potential for increasing the area under cultivation and, thereby, decreasing further the need for imported high-protein animal feedstuff. There is also an opportunity to widen the market for pea seeds as a European source of starch and protein for use by food processors and by the chemical industry.

Comparison between cereal, potato and pea starches

The most widely used starches are extracted from maize seeds. Maize is popular with food processors because, in general, there is no supply problem and genetic variation in this species has given rise to a range of maize starches with amylopectin contents ranging from 20 to 75%. There is an added advantage that the by-products from extracting starch – oil and protein – are also high-value products. The main problems with maize and other cereal starches are technical in that they are relatively difficult to extract. They also contain significant amounts of lipid that adversely affect the functional properties of the starch and which may be oxidized to give the starch 'off-flavours' [2].

* To whom correspondence should be addressed.

In general, potato starches have better functional properties than those from cereals, in particular they have excellent pasting, film-making and binding characteristics [3]. Potato starches also have much lower lipid levels than cereals. Like cereals, the main problems with producing starch from potatoes are technical and concerned with the harvesting, transporting and storing of tubers, which are composed of 80% water. Unlike cereals, the by-products from the potato starch extraction process have little commercial value.

The functional properties of pea starches are characterized by a higher level of solubilization and a slightly restricted swelling of the grains compared with cereal and potato starches [4–6]. The pastes produced by pea starches are consistent and comparable with starches from other sources. The viscosity of heated suspensions, however, is reported to be high when cooled, and stable during storage under high temperature [4–7]. In general, there are no problems in extracting starch from pea seeds and there is the added advantage that the main by-product, protein, has a high commercial value.

For pea to become a serious alternative to cereals and potato as a source of starch for industrial and food uses in Europe, pea starch will either need to be produced more economically than existing starches, or it will need to have properties that are superior or different to those already being marketed. The economics of starch production are largely subject to world commodity prices and to political decisions on world trade and will not be considered here. The present article will concentrate, therefore, on the range and properties of pea starches that are available and those that will become available in the future, making comparisons, where appropriate, to cereal and potato starches that are already in the market place.

Utilizing existing pea mutants (*r* and *rb*) for starch

It is likely that the functional properties of starch are related to the chemical composition and physical structure of the starch grains and that these in turn are in some way determined by the biochemistry of starch synthesis and the genes encoding the enzymes at each step in this process. When the links between these different areas become clearer it should be possible to relate starch biosynthesis to the functional properties of starch. Over the past few years a great deal of information has been accumulated about the biochemistry of starch synthesis in the pea embryo (for a review see [8]), and a series of mutants has been isolated that produce starches that differ in their chemical composition and physicochemical properties [9,10]. The availability of this biochemical information and seed material makes the pea an ideal model system for studying the link between genetic changes in starch synthesis and the functional properties and uses of starch. It may also allow the pea seed to be developed as a source of starches, either for new industrial uses, or as replacements for potato and maize starch.

Effects on starch content and composition

Until the early 1960s, the only mutation known to affect the starch content and composition of pea seeds was at the *r* locus. This mutation caused the dry seed to be wrinkled and was therefore called *rugosus* (from the Latin for wrinkled), which was then abbreviated to *r*. The presence of the *r* mutation reduced the starch content to about 35% of the seed dry weight compared with about 55% for the wild-type seed. It is now known that the mutation is in the gene encoding starch branching enzyme I [11], the activity of this enzyme being absent from seed of the mutant [12,13]. As well as decreasing the starch content, the mutation gave rise to starch that had a greatly reduced level of amylopectin (about 35% compared with about 65% in the wild type). In addition to altering the content and composition of the starch, the *r* mutation affected the shape of the starch grain, which was changed from a simple oval shape to one that was deeply fissured [9,10].

In 1962 Kooistra [14] identified a mutation at a second locus that gave rise to a wrinkled dry seed. This second *rugosus* mutation was called *rb* and was later found to decrease the activity of the enzyme ADP glucose pyrophosphorylase by about 90% [8]. As with the *r* mutation, the presence of this mutation decreased the starch content to about 35% of the seed dry weight. Unlike *r*, however, the *rb* mutation was found to increase the proportion of amylopectin in the starch to about 75% [9,10]. The presence of this mutation has no apparent effect on the shape of the starch granule although there is evidence that it reduces the average starch grain size [15].

Effects on starch structure and physicochemical properties

All starches are composed of amylose, an essentially linear polymer composed of α-(1→4) glycosidic linkages, and amylopectin, a highly branched macromolecular glucose polymer consisting of α-(1→6) branch points in addition to the α-(1→4) links between glucose monomers. When amylopectin is treated with isoamylase, which cleaves the α-(1→6) links, the branches are released and can be separated according to length using chromatography. It has been shown that the branch-length distribution profiles vary in different starches [16]. The presence of the *rb* mutation has no significant effect on the profile but the amylopectin produced by the *r* mutant has a greater average chain length than that from the wild type [15].

Starch grains contain semi-crystalline and disordered, or amorphous, regions. The former are produced by the ordered packing of double helices, formed from short amylopectin chains. Pea starches differ from cereal and potato starches in the way that these crystalline regions are constructed and packed within the starch grain. Starches are composed of two types of polymorph structures, A and B, with the A-type being more dense than B [17,18]. Maize starch is composed of A-type polymorphs and the starch is called A-type, while potato starch has B-type polymorphs and is called B-type. Pea starches are characterized by having both A- and B-type polymorphs within their crystallites and are called C-type starches. The differences between A-, B- and C-type starches can be demonstrated using wide-angle X-ray diffraction, which shows differences in the position and intensity of the peaks for the three types of starches [17].

A comparative study between starches from the wild type and the *r* and *rb* mutants using wide-angle X-ray diffraction showed that all three starches had typical patterns expected for C-type starches (Figure 1). There were differences between the three starches, however, with regard to the relative intensity of the peaks [19]. The most significant effect was attributed to the presence of the mutant

Figure 1

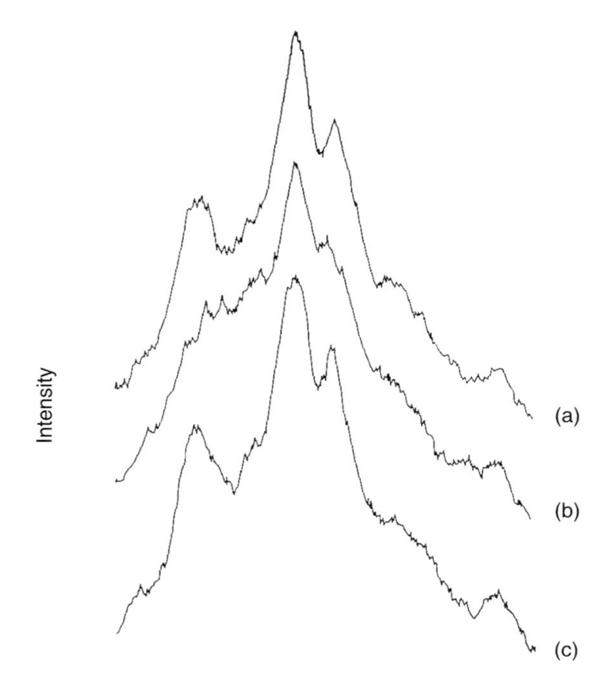

Wide-angle X-ray diffraction patterns for pea starches

(a) Wild-type, (b) rrRbRb and (c) RRrbrb.

r gene, which resulted in a significant decrease in the relative intensity of the peaks at $2\theta = 15.3$ and at 23.5, compared with the wild type, indicating that there were some changes in the sizes of the crystalline lamellae. There was also a significant increase in the relative intensity of the peak at $2\theta = 5.7$, indicating a significant increase in the B polymorph content of this starch.

The structure of the ordered parts of the starch granule varies between the different starches. Using solid-state NMR it is possible to determine the proportion of double helical structures in the starch and this can then be related to the proportion of the starch that is crystalline. In pea and potato starches this technique has demonstrated that a proportion of the double helices are not incorporated into crystalline structures, while in maize this difference is less apparent (Table 1). Starch from the *r* mutant has a very low content of double helices and most of these are organized into crystallites (Table 1).

When starch is suspended in water at room temperature the amorphous phase of the starch granules swells to some degree. During heating this swelling increases and becomes irreversible above a certain temperature. At the same time, the crystalline structure of the starch granules melts cooperatively, a process known as gelatinization. Differential scanning calorimetry (DSC) has been used to study this process in starch from the wild-type and from the *r* and *rb* mutants under quasi-equilibrium conditions (temperature increase of 1 $K \cdot min^{-1}$ and starch concentrations of not more than 2%) [19]. Under these conditions, starch from the wild type and *rb* mutant gave a single, or slightly double, narrow endothermic

Table 1

	Type of starch	Total crystallinity (%)	Double helices (%)	Rigid component (%) 25°C	40°C
Wild type	C	21	37	74	59
r Mutant	C	19	21*	49	28
rb Mutant	C	29	45	72	56
Potato	B	25	53	73	58
Maize	A	39†	42†	68	65

**Estimated value.*

*†Data from Nara, S. (1978) Starch **30**, 183–186.*

Proportion of double helices and total crystallinity, and changes in the rigid component with temperature of different starches

gelatinization peak (Figure 2). Starch from the *r* mutant, however, showed no sharp change in heat capacity, indicating that there was no co-operative transition for this starch under these conditions.

The presence of the *rb* mutation does not significantly affect the gelatinization temperature of the starch, but does have a significant effect on the enthalpy of transition, which is very high for this starch compared with the wild type. It has been suggested that this effect is due to specific charge interactions within the starch granule, most likely between the crystalline and intercrystalline material [19].

Figure 2

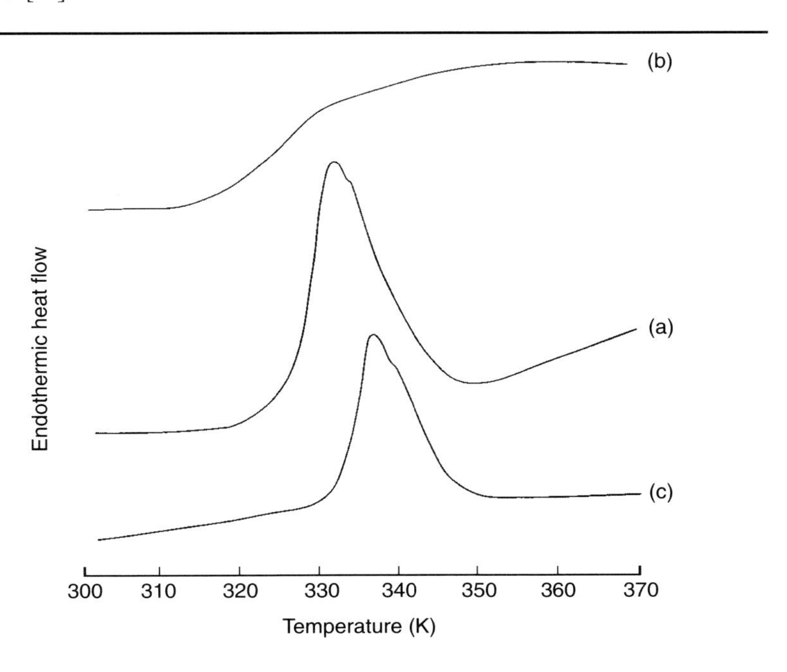

DSC thermograms for pea starches in excess water with a heating rate of 1 K·min^{-1}

The concentration of starch suspension is in parenthesis. (a) Wild type (0.54%), (b) rrRbRb (0.50%) and (c) RRrbrb (0.07%).

When the starches from the wild type and the *rb* mutant are gelatinized in dilute KCl solutions, the DSC curves for both are bifurcated. It has been suggested that the higher temperature peak is due to the melting of A polymorphs and the lower temperature peak to the melting of B polymorphs, these events corresponding to two different co-operative transitions that are independent [19]. This hypothesis has been supported by an experiment during which wild-type pea starch was heated until just after the first transition and then cooled before being reheated. Wide-angle X-ray crystallography demonstrated that the cooled starch attained the characteristics of A-type starch. On heating the second time, this starch gave only a single peak of transition, corresponding to the melting temperature of A polymorphs.

Potential for developing new starches

Isolation of new mutants (*rug3*, *rug4*, *rug5* and *lam*)

Until 1987, the only mutations known to affect the content and composition of starch in pea seeds were those described above. During that year we initiated a programme to produce additional mutants based on chemical mutagenesis. Since both the *r* and *rb* mutations produced seed that had a wrinkled appearance this character was used as a primary screen for new mutants. This initial screen was followed by a chemical analysis of the selected seed to determine the starch and amylose contents. At the end of this screening process about 30 lines were isolated that had wrinkled dry seed [20]. A diallelic crossing programme between the different mutant lines separated the mutants into five complementation groups, indicating that there were mutations at five different genetic loci [21]. There were a number of lines within each complementation group, indicating that a number of mutant alleles had been produced at each locus. In addition, the presence of multiple alleles at each locus indicated that the screen covered all loci that result in a wrinkled seed. Two of the complementation groups contained mutants that were allelic to either the original *r* or *rb* mutant alleles and were assigned, therefore, to either the *r* or *rb* loci. The other three groups were at previously unknown loci and were termed *rug3*, *rug4* and *rug5*, using the contemporary nomenclature for the rugosus character [21].

The selection of all of the rugosus mutants depended on a change in shape of the seed, which is now known to be associated invariably with a decrease in the content of starch in the embryo. Any mutation that affected starch composition without reducing the starch content, however, would probably produce a round seed and would not have been selected using the seed shape screen. Such mutants are known in other species; for example, waxy maize has a normal content of starch that has a very low amylose content, and a normal seed shape [22]. The selection of such mutants required a different type of screen. In the case of low-amylose mutants it was based on the staining properties of starch with iodine. Amylose-free (*amf*) mutants of potato had been selected by painting iodine solution on to the cut surface of potato tubers [23]. The mutants stained a red–brown colour rather than the indigo colour normally associated with starches containing the more usual 30% amylose. A method was devised for screening the mutant lines from the mutagenesis programme based on this colour change [24]. This entailed transferring some of the starch grains from the seeds on to filters and then staining the filters with iodine solution. The screen identified five mutant lines, all of which had seeds that were similar in shape to the wild type and starches that stained red–brown with iodine. Genetic analysis of the new mutants has shown that they are allelic monogenic recessives. They have been assigned the symbol *lam* (*low amylose*) [24].

Starch and starch grain characteristics

The effect of mutant alleles at either the *r* or *rb* locus on starch content and composition has been discussed earlier. Mutations at the three new *rugosus* loci also have dramatic effects on the content and composition of the starch (Table 2) [9,10]. In general, the *rug3* mutants have very low levels of starch in their seeds. One line, however, ($rug3^a$), has about 12% starch present and in this mutant the amylose content of the starch is also relatively low (approx. 12%). The starch contents of both the *rug4* and *rug5* mutants are also reduced, the effect being more significant in *rug5* lines than *rug4* (Table 2). The starches from *rug4* mutants, however, have a lower proportion of amylose than those from *rug5* (Table 2). The starch grains from *rug5* mutants have a unique, irregular, appearance that has not been reported for any other starch from seeds. There is also some evidence that the amylopectin isolated from this starch may be very different from the wild type in that, following debranching, some of the side chains appear to be very long [15]. It is now known that the presence of mutant genes at the *rug4* locus results in a decrease in the activity of sucrose synthase [25] and that *rug5* mutants lack the activity of a starch synthase [25], which is thought to contribute most of the activity in the wild-type pea seed [26].

As expected, the starch contents of the *lam* mutants were similar to that found in the wild type and the amylose contents were very low or undetectable, depending on the method of analysis. The starch grains from the *lam* mutants appear similar in shape to those from the wild type but on staining with iodine they attain a blue core with a pale periphery. A biochemical analysis has shown that the *lam* mutants lack granule-bound starch synthase I activity [24].

The discovery of the new mutants raised the number of loci known to affect starch content and composition in pea seeds to six. In addition to the major effects attributed to differences between these loci, there are small effects between the mutant alleles within each locus. For example, a detailed study of mutants at

Table 2

	Starch (% seed d. wt.)	Amylose (% starch)
Wild-type	50	35
r	27–36	60–75
rb	30–37	23–32
rug 3	1–12	12
rug 4	38–43	31–33
rug 5	29–36	43–52
lam	39–49	4–10

Range of variation for starch and amylose contents in seeds of pea mutants

the *r* locus has shown that one is a null variant and produces no starch branching enzyme I protein [27]. Seed of this mutant has the lowest starch content. However, the presence of another of the mutant *r* alleles has a much weaker effect on the starch content and this relates to the fact that the mutation is in one of the least conserved regions of the protein [27].

Structure and physicochemical properties

Studies on the structural and physicochemical properties of the starches from the new mutants are still at an early stage and are based on the information gained from analysing the wild-type, *r* and *rb* mutant lines discussed earlier. From wide-angle X-ray diffraction studies of all the starches, it is apparent that there is a wide range of variation with regard to relative peak height, but all mutant starches had the same peak position, establishing that they were all C-type starches, typical of grain legumes (T. Bogracheva, unpublished work). With regard to gelatinization, the new starches are either similar in behaviour to wild-type and *rb* starch, in that they show a narrow cooperative peak of gelatinization, or they are similar to the *r*-type starch, with no discernable peak of transition (T. Bogracheva, unpublished work). Starch from the *rug3, rug4* and *lam* mutants fell into the first group and *rug5* starch could be placed in the second. Within the first group of starches (wild-type, *rb, rug3, rug4* and *lam*) there was a 10 K range in peak temperatures. It was also apparent that there was no direct relationship between the amylose content, the content of A and B polymorphs and the thermodynamic parameters of gelatinization. The lack of a co-operative peak of transition in the second group (*r* and *rug5*) indicates that they cannot be referred to as first-order transitions.

NMR relaxation has identified two components in starch granules, rigid and mobile. The rigid component is believed to be composed mainly of ordered structures coupled with part of the amorphous material. The remaining part of the amorphous material comprises the mobile component. It is thought that during heating in excess water the rigid component in the starch decreases because of decoupling of the ordered and amorphous parts. The rate of decrease of the rigid component during heating in excess of water differed between the two groups of starches (Table 1; T. Bogracheva, unpublished work). The rigid components for the second group of starches were rapidly reduced and disappeared after heating to 40 K, while the rigid components from the first group decreased much more slowly.

The differences in behaviour between the two starch groups during heating was reflected in structural differences between the crystalline part of the amylopectins. This part of the amylopectin was debranched and the distribution of branch sizes was determined using ion-exchange chromatography. The distribution patterns for the first group of starches showed single peak curves, similar to those obtained for maize and potato. The patterns for the second group, however, contained multiple peaks with a large proportion of chains having a higher degree of polymerization.

Creation and characterization of double mutants

The genetic characterization of the pea mutants has shown that they can be divided into six complementation groups, or loci, each containing a number of different mutant alleles. It is evident that the presence of a single mutation from any of these groups gives rise to a starch that differs from that of the wild type. Having determined that the six loci are genetically independent it becomes feasible to link the mutations in pairs or even in groups of three or more, to give potentially an even wider range of starches. We are in the process of combining the mutations in pairs with the objective of studying the properties of the resulting starches. Such an exercise will also give us information on the relationship between the loci with regard to epistasis or gene interaction.

We already have a great deal of information about the double mutant created by combining the original *r* and *rb* mutations [9,10,19]. The starch content of the *r/rb* mutant is reduced to about 23%, which is less than that of seed containing either of the single mutations. The amylose content of the *r/rb* starch is about 49%, which is intermediate between that found in the starch from the two single mutant lines. The shape of the starch grains is compound-like, which is similar to that found in starch from the *r* mutant. The starch granules, however, are on average smaller than those found in starch from the single mutant. The physicochemical properties of the *r/rb* starch differ from those observed when either of the single mutations was present. With regard to the wide-angle X-ray pattern, the relative intensities of the peaks at $2\theta = 15.3$ and $23.5°$ were smaller than for the wild type, indicating that there were changes in the crystalline lamellae parameters, although these changes were less than when only the *r* mutation was present. The B-polymorph content of the *r/rb* starch was about 92%, which was higher than that found in either the *r*, *rb* or wild-type starches. During gelatinization in excess water, the *r/rb* starch gave a single very broad peak at a higher temperature than that for either the wild-type or *rb* starch. In addition, unlike the wild-type and *rb* starches, only a single peak remained even in salt solutions. Overall, it was evident that the crystalline structure of the double mutant starch reflected the contrasting influence of both mutant genes.

Although we are only at the beginning of our studies of starches from other paired mutant combinations, the information to date suggests that the starches will be of great interest. We have found that combinations that include genes known to affect starch grain shape, for example those including *rug5*, give rise to even more extreme grain shapes [15]. Any paired combination of genes from the five *rugosus* loci always gives rise to amounts of starch in the seed that are lower than either parent and, if one of the *r* mutants is involved, then the amylose content of the resulting starch is always higher than that from the wild type. When the *lam* mutants are combined with the *rugosus* mutants the resulting starches always contain an amylose fraction, as determined by the iodine staining properties (T. Wang and L. Barber, unpublished work). A similar amylose fraction is produced in maize seeds when the *wx* mutations, which are analogous to *lam*, are combined with other mutations affecting starch synthesis. This is apparently the result of an anomalous amylopectin with longer chain lengths than those found in the non-mutant amylopectin [22].

Will peas become a viable starch crop?

The holistic approach

The pea crop within the U.K. and much of N. Europe is used mainly for compounding into animal feedstuff. A single mutant gene (*r*) has created an additional use of pea seeds, as a vegetable following harvesting when immature and then either freezing, quick-drying or canning. The use of the crop as a source of components for the food or chemical industry, however, has never been developed within Europe. The question is whether such a development is feasible or even desirable? The answer in both cases is undoubtedly yes. The wild-type pea is composed of about 55% starch, 22% protein and 2% oil, plus other minor components such as soluble carbohydrates and 'fibre'. All of the seed components can be separated relatively easily and this is already being carried out commercially within the E.U. [28]. There is an expanding market for pea protein to be processed as an alternative to meat products, and for pea fibre, which has very good properties for improving the texture of processed food products. The development of the pea as a starch crop, therefore, has to be viewed in the light of the usefulness and markets for these 'by-products'. As stated earlier, pea starch is unique and quite different in properties from maize and potato starches. By increasing the available variation and learning more about these properties we will increase the potential uses.

Designer starches

It has been shown recently that high-amylose starches from the wrinkled (*r* allele) pea have improved mechanical properties in bioplastics [29]. At present, the *r* mutant pea is the only crop containing high-amylose starch that can be grown within the E.U. This example represents the effect of a single mutation giving rise to a starch with specific properties that can be utilized for an industrial purpose, in this case bioplastics. Within this paper we have discussed the development of a range of mutants, which, even from our preliminary analysis, have shown a range of properties. Our aim is to link an understanding of the physicochemical properties of the starches from pea to functional properties, in particular those concerned with rheology, which reflect many of the uses within industry. In addition, we are also trying to link genetic changes in the biosynthesis of starch to its chemical structure and physicochemical properties. If we can achieve the necessary understanding at each step in this interdisciplinary chain, then it should be possible to manipulate the genetic constitution of a pea seed with the knowledge that the resulting starch will have specific properties that can be utilized within a specific industrial process. At present, 'designer starches' are still just a concept, but we are confident that the unique mutant material we are developing in pea, coupled with the necessary interdisciplinary research that we are putting together, will bring this idea to fruition in the not too distant future.

References

1. Carrouee, B. (1995) Proceedings of the 2nd European Conference on Grain Legumes, July 1995, Copenhagen, pp. 2–3, AEP, Paris
2. Morrison, W.R. and Gadan, H. (1987) J. Cereal Sci. **5**, 263–275
3. Mitch, E.L. (1984) in Starch: Chemistry and Technology (Whistler, R.L., Bemiller, J.N. and Paschall, E.F., eds.), pp. 479–490, Academic Press, London
4. Doublier, J.L. (1987) J. Cereal Sci. **5**, 247–262
5. Hoover, R. and Vasanthan, T. (1994) Carbohydr. Res. **252**, 33–53
6. Davydova, N.I., Leontiev, S.P., Genin, Ya.V., Sasov, A.Yu. and Bogracheva, T.Ya. (1995) Carbohydr. Polym. **27**, 109–115
7. Stute, R. (1990) Starch/Starke **42**, 178–184
8. Smith, A.M. and Denyer, K. (1992) New Phytol. **122**, 21–33
9. Wang, T.L. and Hedley, C.L. (1993) in Peas: Genetics, Molecular Biology and Biotechnology (Casey, R. and Davies, D.R., eds.), pp. 83–120, CAB International Press, Wallingford, U.K.
10. Wang, T.L. and Hedley, C.L. (1991) Seed Sci. Res. **1**, 3–14
11. Martin, C. and Smith, A.M. (1995) Plant Cell **7**, 971–985
12. Smith. A.M. (1988) Planta **175**, 270–279
13. Bhattacharyya, M.K., Smith, A.M., Ellis, T.H.N., Hedley, C. and Martin, C. (1990) Cell **60**, 115–121
14. Kooistra, E. (1962) Euphytica **11**, 357–373
15. Lloyd, J.R. (1995) PhD Thesis, University of East Anglia
16. Hizukuri, S. (1986) Carbohydr. Res. **147**, 342–347
17. Sarco, A. and Wu, H-Ch. (1978) Starch/Starke **30**, 73–78
18. Imberty, A. and Perez, S. (1988) Biopolymers **27**, 1205–1221
19. Bogracheva, T.Ya., Davydova, N.I., Genin, Ya.V. and Hedley, C.L. (1995) J. Exp. Bot. **46**, 1905–1913
20. Wang, T.L., Hadavizideh, A., Harwood, A., Welham, T.J., Harwood, W.A., Faulks, R. and Hedley, C.L. (1990) Plant Breed. **105**, 311–320
21. Wang, T.L. and Hedley, C.L. (1993) Pisum Genet. **25**, 64–70
22. Nelson, O. and Pan, D. (1995) Annu. Rev. Plant Physiol. Plant Mol. Biol. **46**, 475–496
23. Hovenkamp-Hermelink, J.H.M., Jacobson, E., Ponstein, A.S., Visser, R.G.F., Vos-Scheperkeuter, G.H., Bijmolt, E.W., de Vries, J.N., Witholt, B. and Feenstra, W.J. (1987) Theor. Appl. Genet. **75**, 217–221
24. Denyer, K., Barber, L.M., Burton, R., Hedley, C.L., Hylton, C.M., Johnson, S., Jones, D.A., Marshall, A.M., Smith, A.M., Tatge, H., Tomlinson, K. and Wang, T.L. (1995) Plant Cell Environ. **18**, 1019–1026
25. Craig, J., Smith, A., Wang, T., Lloyd, J. and Hedley, C. (1995) Proceedings of the 2nd European Conference on Grain Legumes, July 1995, Copenhagen, p. 396, AEP, Paris
26. Smith, A.M. (1990) Planta **182**, 599–604
27. MacLeod, M. (1994) PhD Thesis, University of East Anglia
28. Hedley, C.L. and Wang, T.L. (1993) Agro. Food Ind. Hi Tech. 14–17
29. Colonna, P., Lourdin, D., Della-Valle, G. and Buleon, A. (1995) Proceeding of the 2nd European Conference on Grain Legumes, July 1995, Copenhagen, pp. 354–355, AEP, Paris

Manipulating the starch composition of potato

A.J. Kortstee, E. Flipse*, A.G.J. Kuipers, E. Jacobsen and R.G.F. Visser†
Graduate School of Experimental Plant Sciences, Department of Plant Breeding, Agricultural University Wageningen, P.O. Box 386, 6700 AJ Wageningen, The Netherlands

Abstract

Starch can be fractionated into two types of glucose polymers: amylose and amylopectin. Amylose consists of essentially linear chains of α-(1,4)-linked glucose residues, whereas amylopectin is built up from α-(1,4)-linked chains with α-(1,6)-linked branches. The composition and fine structure of starch are responsible for many of the physicochemical properties and thus determines its industrial uses. Variation in starch structure and composition can be found between and within crops. In the latter case it can be found in mutants, often resulting from the loss of function of one or more of the genes involved in starch biosynthesis. In maize, the most extensively studied crop, mutant genotypes are known for nearly every gene identified as being involved in starch biosynthesis. Differences in starch compositon can also be achieved by genetic modifications such as antisense inhibition of genes or overexpression of (heterologous) genes. Most examples of genetic modification of starch composition are in potato, which can easily be transformed. Antisense inhibition of enzymes in the biosynthetic pathway, such as ADP glucose phosphorylase (AGP), (granule-bound) starch synthase or branching enzyme, lead to an altered starch content and/or composition. In addition, the introduction and expression of bacterial genes, such as genes of the *Escherichia coli* glycogen synthesis pathway, in potato leads to starches with altered content, composition, structure and physicochemical properties. Studying the physicochemical properties of these altered starches will, together with the information obtained by research on starches of mutants, help to clarify the precise relationship between structural and functional features of starch.

Introduction

Starch biosynthesis and structure

In higher plants the major reserve polysaccharide is starch. Starch is stored in the form of granules in special starch-storing plastids, the amyloplasts. Synthesis of starch in the amyloplast involves the combined actions of a number of enzymes,

*Present address: Scottish Crop Research Institute, Invergowrie, Dundee DD2 5DA, Scotland, U.K.
†To whom correspondence should be addressed.

starting with AGP, which synthesizes ADP-glucose. In the next step the glucosyl part of ADP-glucose is linked to the non-reducing end of a pre-existing glucan chain by a starch synthase (SS), forming an α-(1,4)-linkage while releasing ADP. Branching of linear α-(1,4)-chains is catalysed by branching enzymes, which hydrolyse an α-(1,4)-linkage within a chain and then form an α-(1,6)-linkage between the reducing end of the chain that was cut, and another glucose residue, probably from the hydrolysed chain [1]. One class of starch synthases, the granule-bound starch synthases (GBSSs), synthesizes an α-(1,4)-linked glucan that is called amylose and usually makes up 15–30% of the total starch. The glucan formed by the synthases and subsequently branched by the branching enzymes is called amylopectin and forms the major part of starch [2]. Amylopectin consists of short linear α-(1,4)-linked chains with an average length of about 25 glucose residues. The polymodal chain length distribution and the clustering of branch points along the axis of the molecule were first described in a model by Hizukuri [3]. In this model of amylopectin, three types of chains can be designated: the A-chains that are connected to the amylopectin by their potential reducing endgroups and the B-chains that are similar to A-chains, but carry one or more A and/or B-chains. The C-chain of the molecule is the chain that carries the sole reducing endgroup [4]. The cluster model for amylopectin is depicted in Figure 1.

Apart from amylose and amylopectin, the starch granule contains protein (of which GBSS is the most prominent one), phosphorus (both starch-bound and free), lipid and some ash [5].

In addition to the two major starch fractions, amylose and amylopectin, some researchers have reported the existence of a third starch fraction, the intermediate fraction [6]. This fraction is evident from comparison of gel-permeation elution profiles of native and debranched starch, and is believed to contain highly branched amylose and long-chained amylopectin [6,7]. However, the existence of the intermediate fraction is still uncertain [8], as other researchers argue that it does not give a distinct peak on gel-permeation analysis of isoamylase debranched starch, as can be seen for amylose and amylopectin.

Figure 1

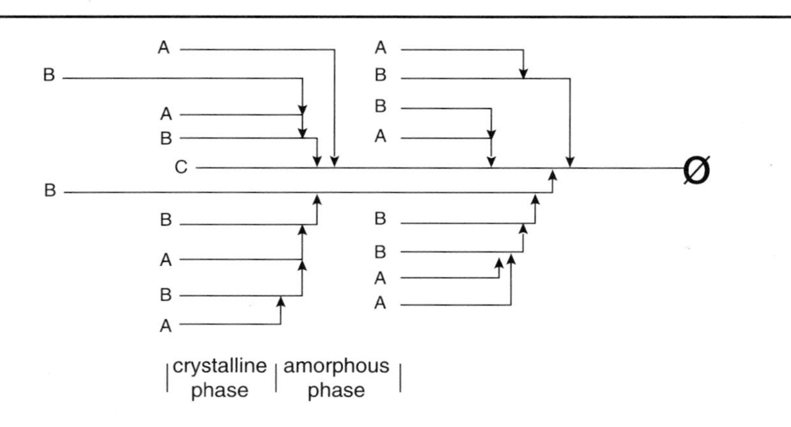

Amylopectin fine structure according to Hizukuri [3]

ϕ, Reducing endgroup; —, α-(1,4)-linked chain; →, α-(1,6)-linkage.

Starch composition

The composition of starch is determined by the amylose content, molecular size of amylose and amylopectin, average side-chain length, chain length distribution and the non-carbohydrate components and is considered to be unique for each botanical source [5]. Genetic variation within species can contribute considerably to the wide variation found in structural characteristics for starch of different origins. This is also the reason for the distinctive properties of native starches from different botanical origins and even from different genotypes within a species [8–11]. The influence of structural characteristics on the physicochemical properties of the starch is evident [5]. The precise nature of this relationship, however, is more difficult to establish. In attempts to define the relationship between the structural and functional features of starch, many studies were undertaken involving native starches of different botanical origin [5,11,12] and mutants with altered starch compositions [10,13–17]. Most mutants are found in maize, where they exist for almost every isoform of the enzymes involved in starch biosynthesis, resulting in starches with different compositions and characteristics [6,13,14]. The best-studied example of mutant starch is that of the *waxy* mutant of maize [18]. This mutation has also been identified and studied in other crops, including potato, rice, barley and wheat (see [19,2,20,21] respectively). These mutants contain amylose-free starch as a result of a defect in the gene encoding GBSS and a subsequent absence of GBSS activity [18]. Another group of mutants contains a relatively high amount of amylose in the starch as a result of a mutation in one of the isoforms of starch branching enzyme and is found in maize (*amylose-extender*) [22,23] and pea (*wrinkled*) [24,25].

Genetic modification of starch content and composition

Another source of alternative starch composition, besides mutants, is the use of genetic modification. This work has mostly been carried out in potato, the most easily transformed of the starch-storing crops. By inhibiting GBSS gene expression via the so-called antisense RNA inhibition method, potato starch was obtained that was completely amylose-free [26,27,41]. The same approach was used to inhibit the branching enzyme gene, resulting in plants with reductions in branching enzyme RNA and protein. However, there were no detectable changes in amylopectin structure, such as the degree of branching, although the physicochemical properties (mainly the viscosity) of the starch were changed [28]. Inhibiting the potato AGP resulted in sugar-storing tubers with a decreased amount of starch and distorted tuber morphology [29], but no structural changes in the starch.

As an alternative to inhibiting gene function, homologous or foreign genes can be introduced to change the starch composition. Re-introduction of GBSS into the amylose-free (*amf*) mutant of potato [30,31] resulted in tubers with up to wild-type level of amylose in the starch. In contrast, inserting extra copies of the same gene into an amylose-containing potato line did not have any effect on the amount of amylose. Indeed, the opposite effect was sometimes observed, with the amount of amylose decreasing after insertion of extra GBSS gene copies [31].

The introduction and expression of cyclodextrin glycosyltransferase (CGTase) from *Klebsiella* sp. resulted in an unique type of starch that contained an appreciable amount of cyclodextrins [32]. The genes from the glycogen synthesis operon of *E. coli*, *glgA*, *B* and *C*, have also all been expressed in potato. Expression of a mutant *glgC* gene, the AGP, in the potato variety Russet Burbank, resulted in an increased amount of starch [33] that displayed no structural changes. Expression of the glycogen synthase gene, *glgA*, led to an altered potato starch composition, with a reduced amylose content and an increased degree of branching in the amylopectin [34]. These altered structural features of the starch also had an effect on the physical properties such as viscosity and gelatinization.

We have also described how the expression of the *E. coli glgB* gene, encoding glycogen branching enzyme, in the *amf* potato mutant resulted in amylopectin with an increased degree of branching [35].

Analysis of the physicochemical properties of starches of transgenic plants, in relation to their altered composition, will now be discussed.

Relationship between structural and functional properties of starch

Analysis of starches of mutant genotypes in four inbred maize lines [10] showed the importance of the ratio of short to longer chains for the thermal behaviour of starch. Starch from the double mutant *aewx* was found to contain more chains of high molecular mass, the population of lower-molecular-mass chains consisted of longer chains and a higher T_{max} compared with starch from the single mutant *wx*. This effect was thought to be caused by an increased proportion of starch in the crystalline form as a result of a higher proportion of B1 chains.

Wang and co-workers [13,14] described the structural characteristics of starches from 17 mutant maize genotypes and their effects on physicochemical behaviour. They concluded that the precise relationship between starch structure and physical properties was not always clear, but found that the amylose content had a large influence on swelling and gelatinization. The amylose content was negatively correlated with the swelling power, light transmittance of starch solutions (%T) and peak viscosity and positively correlated with Blue Value and λ_{max}. Other properties such as the intermediate size content and the ratio of short to longer chains were found to be negatively correlated with peak viscosity.

Jane and Chen [36] conducted a study of the effects of the molecular size of amylose and the average side-chain length of amylopectin on the paste properties of rice and maize starches. Rice amylopectin was found to have both higher viscosity and shorter chain length compared with both high-amylose maize and waxy maize, which contrasts with the results obtained by Wang and co-workers [13]. However, the phosphorus content of the rice amylopectin was much higher and this could have contributed to the higher viscosity in addition to other structural characteristics of rice amylopectin, which also differed from those of the maize amylopectin.

Influence of amylose in transgenic potato

The ability of starch granules to take up water during heating is accompanied by swelling of the granule. Experiments with normal and waxy cereal starches [37] showed that swelling is an exclusive property of the amylopectin and it involves the whole amylopectin molecule. Furthermore, in normal starches, amylose and lipids can actively inhibit swelling.

Starch from transgenic potato plants with partly inhibited GBSS activity and a lowered amylose content were found to have a distinct swelling pattern [38]. Compared with the untransformed control with a normal amylose content the starch with the decreased amount of amylose (about half of the wild-type level) began to swell at a higher temperature (60 instead of 58 °C), but swelled faster as the temperature increased further. At 65 °C the swelling power of the lower amylose-containing starches were twice as high as the swelling power of the control, and at 70 °C it was approx. 150% of the control value.

Starch samples from the above-mentioned antisense GBSS-inhibited plants were analysed by differential scanning calorimetry (DSC) to determine their gelatinization parameters. Starch from the transgenic plants with a lowered amylose content displayed a higher temperature of onset of gelatinization (T_o), and consequently a higher peak-temperature (T_p) and termination-temperature (T_t) [38]. These results were confirmed by those of Flipse and co-workers [39] who described how the re-introduction of GBSS into an *amf* mutant of potato restored the amylose content to wild-type level. DSC analysis of the starch showed a lowering of T_o for starches with an increased amount of amylose. Likewise, the dynamic rheological properties of this type of starch, determined by a Bohlin VOR Rheometer, again showed the influence of amylose. The changes in storage moduli (G') of 5% starch suspensions during heating and cooling of amylose-containing starches showed sharp increases, indicating that gelatinization was occurring at a lower temperature than in the amylose-free control [39]. In addition, the peak-moduli for the amylose-containing starches were much higher compared with the amylose-free control, and the increase in moduli after cooling, as a result of retrogradation of the amylose, was clearly present in the amylose-containing starches and absent for the amylose-free control.

Antisense GBSS-inhibited plants containing starch with a decreased amount of amylose were also subjected to Bohlin VOR Rheometer analysis, and showed similar results concerning the temperature at which a sharp increase of the storage modulus (G') occurred with respect to the amylose content [38]. With the low-amylose starch the increase in storage modulus occurred at a higher temperature than with the normal amylose control.

In general, the height of the peak modulus (G') and the increase in modulus after cooling were related to the presence or absence of amylose rather than to the amount of amylose.

Relative compression studies with 10 and 20% potato starch gels using an ETIA Tec T04 texture analyser showed the influence of the presence of amylose on the gel strength. Gels made with starch from an amylose-free potato clone were very weak compared with those prepared from amylose-containing starches. In fact it was difficult to even break the gels formed with 10% amylose-free starch gel, requiring twice the force necessary to break amylose-containing gels. Starch gels from a clone with a lowered amylose content (about half of the normal amount) had a gel strength between those of the amylose-containing and the amylose-free starch gels [38].

Influence of branching degree

Flipse and co-workers [28] introduced the endogenous branching enzyme into potato in the antisense orientation in an attempt to decrease the degree of branching of the amylopectin by antisense inhibition. In an amylose-free potato clone, in which the starch granules stained red with iodine, the expression of branching enzyme at the mRNA and protein levels was partly or fully inhibited. The resulting transgenic plants contained starch granules with a small blueish staining core and a large red-staining outer layer, suggesting the presence in the core of a long chain or loosely branched glucan. Although measurements of the amylose content, branching degree or λ_{max} of the starch, revealed no differences compared with the untransformed control, the Bohlin pasting profile of this type of starch showed an increased storage modulus (G') and an increased peak-viscosity compared with the control [28].

Transgenic potatoes were studied that contained starch with an increased degree of branching of the amylopectin as a result of the expression of the glycogen branching enzyme gene (*glgB*) of *Anacystis nidulans* or *E. coli*. These transgenes were expressed in the normal amylose-containing (wild type) background and in an amylose-free (*amf*) potato mutant, respectively. The degree of branching of these starches [expressed as dextrose equivalent (DE), after isoamylase digestion] was increased by up to 25%. This increase in branching degree could be partly explained by the presence of 5 to 15% more short chains in the amylopectin, the so-called A chains. The influence of the altered branching degree on the physicochemical properties of the starches was investigated. No change in granule size or granule morphology was observed for the altered starches of these transgenic plants. However, regardless of the presence or absence of amylose, starches with an increased branching degree (up to 25% higher) showed a lower peak viscosity when suspensions were heated, and (for the amylose-free starches with an increased branching degree) a tendency to form weaker gels (A.J. Kortstee, L.C.J.M. Suurs, A.M.G. Vermeesch, C.J.A.M. Keetels, E. Jacobsen and R.G.F. Visser, unpublished work).

After introduction of the *E. coli glgA* gene, encoding glycogen synthase, into potato a change in the starch composition could be observed. An increased branching degree (as measured by HPLC analysis of isoamylase debranched starch) and a decreased amylose content were described for starch of *glgA*-expressing potatoes [34]. Some physical properties of the starch, such as gelatinization and thermal behaviour, were also changed. The value of T_o measured by DSC was lower and rapid visco analysis showed a higher paste temperature and a lower peak-viscosity for the starch of the transgenic potatoes. The higher paste temperature could possibly be attributed to the lowered amylose content as could the lowered peak viscosity. The lowered T_o, however, could not be attributed to the decrease in amylose content and may have resulted from the increased branching from amylopectin or from a change in another structural characteristic. The much reduced starch content and increased sugar content of tubers indicated a severe disturbance of starch biosynthesis in these transgenic tubers.

Discussion

Table 1 gives examples of the effects of genetic modification on potato starch composition. Effects on the starch composition range from those affecting a single trait, e.g. amylose content, to severe disturbance of the pathway of starch biosynthesis and tuber development. Severe reduction of the major soluble starch synthase SSS III, which was recently identified, seemed to have no influence on the amylopectin : amylose ratio nor on the total starch content but did affect the granule shape. Two types of granules were present in the starch from transgenic tubers: simple granules with deep, often T-shaped, cracks centred on the hilum, and granules that appeared to be large clusters of tiny, spherical granules [40].

Relationship between structural and functional properties

In potato starch as in maize, the amylose fraction is most important in determining the physicochemical properties. The presence of amylose in the starch decreases the swelling power and also decreases the gelatinization temperature. An explanation could be that reduction in the amylose content leads to relatively more crystalline material being present in the granule, resulting in an increase in T_o for DSC analysis. Amylose-free starch gels were very weak, probably because of the lack of amylose to form a network structure [11], so the gel strength depends on the stiffness of the granules themselves. The effect of amylose on the pasting properties was to lower the gelatinization temperature and increase the viscosity after cooling, an effect known as retrogradation. In starch suspensions with a concentration of 5% or more the peak viscosity of amylose-free potato starches is much lower compared with amylose-containing potato

Table 1

Gene		
Antisense inhibition	Effect	Reference
gbss I	Amylose-free starch	[26,27,41]
be	Blue staining core (in *amf* granules)	[28]
sss III	Compound granules	[40]
agpase	Sugar-storing tubers, distorted tuber formation	[29]
Sense (over)expression		
gbss potato	Restored amylose content in *amf*-mutant	[30,31]
glgA	Less amylose, higher-branching-degree amylopectin Disturbed starch biosynthesis	[34]
glgB	Higher-branching-degree amylopectin	[35]
glgC	Increased amount of starch	[33]
cgtase	Cyclodextrin production	[32]

Genetic modification of potato starch

Genes in table: agpase, ADP-glucose pyrophosphorylase; be, branching enzyme; gbss I, granule-bound starch synthase; sss III, soluble starch synthase III; glgA,-B and -C, genes from the glycogen synthesis operon of E. coli; cgtase, cyclodextrin glycosyltransferase gene from Klebsiella sp.

starches. For maize starch both similar and opposite differences to those observed in potato were found between amylose-free starch and amylose-containing starch, depending on the starch concentration and the bowl speed of the visco-amylograph.

Increasing the degree of branching of the starch by introduction and expression of a bacterial glycogen branching enzyme had no dramatic effect on the physicochemical behaviour of the starch, but small differences were detected between starch with a 25% increased degree of branching and its untransformed control. Analysis of changes in the storage modulus (G') of 5% starch suspensions showed a decreased peak-viscosity for starch with an increased degree of branching. This was supported by the results of experiments in which the expression of endogenous branching enzyme was inhibited. This type of starch, which presumably had more longer chains, showed an increase in peak viscosity when compared with control material (as can be seen in Figure 2).

An increase in shorter chains therefore leads to a decrease in peak viscosity and an increase in longer chains leads to increased peak viscosity. These results with transgenic potato are in accordance with the results obtained by Wang and co-workers on maize mutants [13,14]. They also find a negative correlation between the ratio of short to longer chains and the peak viscosity of starch suspensions. Correlations between the amount of shorter chains and other properties were less clear, but there were indications that starches with an increased degree of branching had decreased gel strength and a higher T_o of gelatinization.

Figure 2

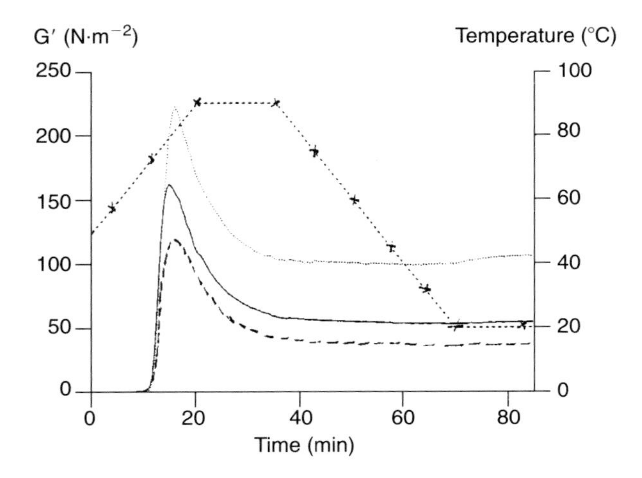

Changes in storage moduli (G') of 5% starch suspensions during heating and subsequent cooling

Temperature against time (······); amylose-free control starch (——); starch from an antisense BE inhibited plant (·····); starch from an E. coli glgB expressing plant with an increased branching degree (- - -).*

In general, it can be concluded that the changes in starch composition in transgenic plants result in specific changes in the physicochemical properties of the starch. The information gained by this type of experiment can, together with information obtained by studies on mutants, help to clarify the relationship between the structural and functional properties of starch. So far our studies of starch from transgenic plants have shown that, although the presence of amylose seems to have the greatest impact on the physicochemical properties, the degree of branching of the amylopectin also plays a distinct role in determining starch characteristics.

References

1. Martin, C. and Smith, A.M. (1995) Plant Cell **7**, 971–985
2. Shannon, J.C. and Garwood, D.L. (1984) in Starch: Chemistry and Technology (Whistler, R.L., BeMiller, J.N. and Paschall, E.F., eds.), 2nd edn., pp. 25–86, Academic Press, Orlando,
3. Hizukuri, S. (1986) Carbohydr. Res. **147**, 342–347
4. Manners, D.J. (1989) Carbohydr. Polym. **11**, 87–112
5. Swinkels, J.J.M. (1985) Starch **37**, 1–5
6. Wang, Y.J., White, P., Pollak, L. and Jane, J. (1993) Cereal Chem. **70**, 171–179
7. Whistler, R.L. and Doane, W.M. (1961) Cereal Chem. **38**, 251–256
8. Tester, R.F. and Karkalas, J. (1996) Cereal Chem. **73**, 271–277
9. Campbell, M.R., Pollak, L.M. and White, P.J. (1995) Cereal Chem. **72**, 281–286
10. Sanders, E.B., Thompson, D.B. and Boyer, C.D. (1990) Cereal Chem. **67**, 594–602
11. Howling, D. (1980) Food Chem. **6**, 51–61
12. Jacobs, H., Eerlingen, R.C., Clauwaert, W. and Delcour, J.A. (1995) Cereal Chem. **72**, 480–487
13. Wang, Y.J., White, P. and Pollak, L. (1993) Cereal Chem. **70**, 199–203
14. Wang, Y.J., White, P. and Pollak, L. (1992) Cereal Chem. **69**, 328–334
15. Campbell, M.R., White, P.J. and Pollak, L. (1995) Cereal Chem. **72**, 389–392
16. Bogracheva, T.Ya., Davydova, N.I., Genin, Ya.V. and Hedley, C.L. (1995) J. Exp. Bot. **46**, 1905–1913
17. Lii, C-Y., Shao, Y-Y. and Tseng, K-H. (1995) Cereal Chem. **72**, 393–400
18. Echt, C.S. and Schwartz, D. (1981) Genetics **99**, 275–284
19. Hovenkamp-Hermelink, J.H.M., Jacobsen, E., Ponstein, A.S., Visser, R.G.F., Vos-Scheperkeuter, G.H., Bijmolt, E.W., de Vries, J.N., Witholt, B. and Feenstra, W.J. (1987) Theor. Appl. Genet. **75**, 217–221
20. Nakamura, T., Yamamori, M., Hirano, H., Hidaka, S. and Nagamine, T. (1995) Mol. Gen. Genet. **248**, 253–259
21. Sano, Y. (1984) Theor. Appl. Genet. **68**, 467–473
22. Boyer, C.D. and Preiss, J. (1981) Plant Physiol. **67**, 1141–1145
23. Hedman, K.D. and Boyer, C.D. (1982) Biochem. Genet. **20**, 483–492
24. Edwards, J., Green, J.H. and apRees, T. (1988) Phytochemistry **27**, 1615–1620
25. Smith, A.M. (1988) Planta **125**, 270–279
26. Visser, R.G.F., Somhorst, I., Kuipers, G.J., Ruys, N.J., Feenstra, W.J. and Jacobsen, E. (1991) Mol. Gen. Genet. **225**, 289–296
27. Kuipers, G.J., Jacobsen, E. and Visser, R.G.F. (1994) Plant Cell **6**, 43–52
28. Flipse, E., Suurs, L., Keetels, C.J.A.M., Kossmann, J., Jacobsen, E. and Visser, R.G.F. (1996) Planta **198**, 340–347
29. Müller-Rober, B., Sonnewald, U. and Wilmitzer, L. (1992) EMBO J. **11**, 1229–1238
30. Leij van der, F.R., Visser, R.G.F., Oosterhaven, K., van der Kop, D.A.M., Jacobsen, E. and Feenstra, W.J. (1991) Theor. Appl. Genet. **82**, 289–295
31. Flipse, E., Straatman-Engelen, I. Kuipers, A.G.J., Jacobsen, E. and Visser, R.G.F. (1996) Plant Mol. Biol. **31**, 731–739
32. Oakes, J.V., Shewmaker, C.K. and Stalker, D.M. (1991) Biotechnology **9**, 982–986
33. Stark, D.M., Timmerman, K.P., Barry, G.F., Preiss, J. and Kishore, G.M. (1992) Science **258**, 287–292
34. Shewmaker, C.K., Boyer, C.D., Wiesenborn, D.P., Thompson, D.P., Boersig, M.R., Oakes, J.V. and Stalker, D.M. (1994) Plant Physiol. **104**, 1159–1166
35. Kortstee, A.J., Vermeesch, A.M.G., de Vries, B.J., Jacobsen, E. and Visser, R.G.F. (1996) Plant J. **10**, 83–90
36. Jane, J. and Chen, J. (1992) Cereal Chem. **69**, 60–65
37. Tester, R.F. and Morrison, W.R. (1990) Cereal Chem. **67**, 551–557

38. Kortstee, A.J., Suurs, L.C.J.M., Vermeesch, A.M.G., Jacobsen, E. and Visser, R.G.F. (1997) in Proceedings of the Conference on Starch: Structure and Function (Frazier, P.J., Richmond, P. and Donald, A.M., eds.), pp. 238–247, Cambridge, April 1996, Royal Society of Chemistry, Cambridge
39. Flipse, E., Keetels, C.J.A.M., Jacobsen, E. and Visser, R.G.F. (1996) Theor. Appl. Genet. **92**, 121–127
40. Marshall, J., Sidebottom, C., Debet, M., Martin, C., Smith, A.M. and Edwards, A. (1996) Plant Cell **8**, 1121–1135
41. Salehuzzaman, S.N.I.M., Jacobsen, E. and Visser, R.G.F. (1993) Plant Mol. Biol. **23**, 947–962

Engineering plant starches by the generation of modified plant biosynthetic enzymes

Thomas W. Okita*, Thomas W. Greene, Mary J. Laughlin, Peter Salamone, Ronald Woodbury, Sang-Bong Choi, Hiroyuki Ito, Halil Kavakli and Kim Stephens
Institute of Biological Chemistry, Washington State University, P.O. Box 646340, Pullman, WA 99164-6340, U.S.A.

Introduction

Starch is an essential storage reserve for many plants. It is composed of two components, amylose and amylopectin. Amylose is mainly a linear array of glucosyl residues arranged with $\alpha(1,4)$-linkages while amylopectin contains linear chains of glucosyl residues that are branched by $\alpha(1,6)$-linkages. Starch exists as transitory starch in leaves and other photosynthetically competent cells and as storage starch in non-photosynthetic storage organs such as tubers and seeds [1]. Transitory starch is composed almost entirely of amylopectin whereas storage starch contains significant amounts of amylose in addition to amylopectin. The differences in the gross structural features between transitory and reserve starch are due to the presence of a specific granule-bound starch synthase in storage tissues [2] and, probably, other factors that have yet to be identified [3].

Because of its availability, storage starch is used in a variety of end-use products and industrial applications. In addition to being used in the food industry, it serves as a lubricant, adhesive and chemical additive in plastic, paper and textile manufacturing [4]. The commercial uses of starch are dependent on the inherent functional properties of the starch granule, which include viscosity and gelation properties. These properties, in turn, are dictated by the size of the starch granules, content of amylose, chain length and frequency of branching in amylopectin, degree of phosphorylation and amounts of protein and lipid associated with the starch granule. Many of these structural properties are influenced by the inherent enzymic properties of enzymes involved both in the biosynthesis and turnover of starch. Starch synthases and branching enzymes exist as multiple activities in the cell. Starch synthases exist only as a granule-bound form, as soluble forms or as both granule and soluble forms [5–7]. Likewise, branching enzymes exist both as soluble and granule-bound forms [5,7,8].

The isoforms of starch synthases and branching enzymes exhibit different catalytic properties. Soluble starch synthase (SSS) exists as two major

*To whom correspondence should be addressed.

forms, types I and II, in developing maize endosperm [5,8–10]. Although the apparent affinity for the substrate ADPglucose is similar for the two SSS, the type I enzyme prefers glucan primers with shorter chains whereas the type II enzyme prefers glucan primers with longer chains. Moreover, the two SSS types differ substantially in their apparent affinities for primer. Likewise, branching enzymes types I and II differ in their affinities for amylose and in their chain length preferences [6,11,12].

Starch structure is also affected by enzyme activities more commonly assigned a degradative role in starch metabolism [1]. The most conspicuous example of the direct involvement of a degradative enzyme activity affecting starch structure is the maize mutant *sugary1*, which lacks granular starch and instead accumulates a water-soluble glucan called phytoglycogen [13]. Biochemical evidence suggests that the genetic defect causes a reduction in debranching enzyme activity [14]. This view has been supported by the recent identification and characterization of the *Su1* gene, which codes for an isoamylase-like enzyme [15]. Similarly, phosphorylase and amylase activities may alter starch structure by affecting the chain length of amylopectin [1].

In addition to enzymes that are directly involved in the biosynthesis and modification of starch structure, starch composition is also affected by enzymes involved in the production of the precursors of starch synthesis [6]. Mutation at the *Rb* locus that codes for the ADPglucose pyrophosphorylase (AGP) large subunit of pea cotyledons not only results in less starch but also in an increased proportion of amylopectin. Similarly, other pea mutants defective in supplying starch precursors also show a similar change in starch quality, indicating that the effect on composition is due to a general reduction in starch synthesis.

Starch composition and quality are markedly influenced by the inherent catalytic activities of enzymes involved in starch biosynthesis and degradation as well as by the enzymic processes that lead to the formation of precursors of starch synthesis. In view of this direct causal relationship, new types of starches can potentially be created by the generation of enzymes with novel catalytic activities. Using AGP as a model, we will discuss our present efforts in generating and identifying mutations that affect the allosteric and catalytic properties of this enzyme. Mutant enzymes with altered allosteric response are potentially useful in enhancing starch biosynthesis in developing sink organs and, in turn, increasing overall plant productivity. Likewise, such an approach can be extended to generate new starch synthases and branching enzymes with novel substrate and catalytic properties and, subsequently, new types of starches.

The generation of mutations in the plant AGP

Complementation of *Escherichia coli glg*C$^-$ by expression of plant AGP cDNAs

AGP is a key regulatory enzyme in the biosynthesis of starch in higher plants, and of glycogen in bacteria [9,10,16]. AGP converts ATP and glucose 1-phosphate (Glc 1-P) into ADPglucose, the activated glucosyl donor that is utilized by an α-glucan synthase activity (starch synthase in plants, glycogen synthase in bacteria) to build the linear α(1,4)-glucosyl chains. Both the plant and bacterial AGPs are allosterically regulated by small effector molecules whose nature reflects the major carbon assimilatory pathways in these organisms. In plants, AGP is allosterically activated by 3-phosphoglycerate (3-PGA) and inhibited by orthophosphate (P_i) [9,10], whereas the bacterial enteric enzyme is allosterically activated by fructose 1,6-bisphosphate (Fruc 1,6-P_2) and inhibited by AMP [16]. Although the bacterial and plant enzymes catalyse the same reaction, they have different oligomeric structures. The bacterial AGP is composed of a single subunit, encoded by the *glg*C gene, which oligomerizes to form a homotetramer [16]. In contrast, the plant enzyme is composed of a pair of related but distinct subunit types, the large subunit (LS) and small subunit (SS), that assemble to form a heterotetrameric enzyme [17].

The heterotetrameric structure of the higher plant enzyme raises questions as to the role of each subunit type in enzyme function. To elucidate the role of each subunit in enzyme catalysis and allosteric regulation, cDNAs for the potato LS and SS were isolated and then cloned on individual plasmids in *E. coli* [18]. When both LS and SS sequences are co-expressed in *E. coli*, significant amounts of AGP activity are readily observed. Moreover, expression of the plant AGP is able to complement a mutation in the bacterial AGP structural gene *glg*C and restore the capacity of these cells to accumulate glycogen [18]. This restoration of glycogen accumulation can be easily scored by subjecting the transformants grown on enriched media to I_2 vapour. The ability to express the potato cDNAs and to complement the *glg*C$^-$ mutation permits a mutational approach to assess the structure–function of the plant AGP in the heterologous *E. coli* host.

Mutagenesis and mutant analysis

To assess the possible role of each subunit type in AGP enzyme function, one of the subunit sequences was subjected to chemical mutagenesis and then transformed into cells containing the other untreated subunit cDNA sequence. Hydroxylamine, which preferentially hydroxylates the amino nitrogen at the C-4 position of cytosine and leads to a G:C → A:T transition [19], was used as the mutagen. The AGP LS expression plasmid was incubated with hydroxylamine for various periods and then transformed into a *glg*C$^-$ strain containing the AGP SS cDNA expression plasmid. Hydroxylamine treatment of the LS for 48 h resulted in a 6-fold decrease in the transformation frequency from 1.5×10^6 to 2.5×10^5 transformants·μg^{-1} DNA (Figure 1). In contrast, the mutational frequency, defined as the fraction of transformants that lost the capacity to accumulate glycogen, as shown by the absence of significant I_2 staining, increased over this period, whereupon 29% of the total transformants were deficient in glycogen accumulation. The majority of the mutants were unable to accumulate glycogen and remained unstained when exposed to I_2 vapour as compared with the dark reddish-purple staining of the *glg*C$^-$ strain expressing the wild-type AGP LS and SS cDNA sequences. Some of the mutants, however, exhibited intermediate to light staining phenotypes, indicating that some low levels of glycogen were accumulated and, in turn, some AGP function had been retained. Based on the results of glycogen staining, the analysis of subunit accumulation by ELISA and the measurable level of *in vitro* AGP enzyme activity, the mutants

Figure 1

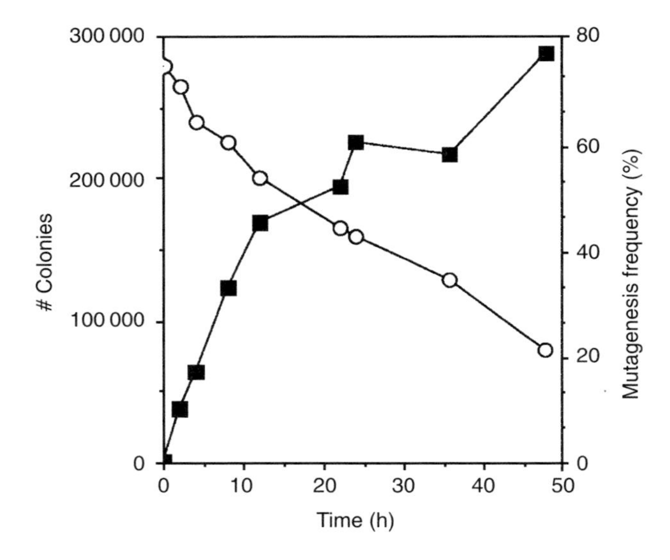

Generation of AGP mutants by hydroxylamine treatment

An expression plasmid containing the LS cDNA was incubated with hydroxylamine as outlined in [20]. DNA samples were periodically removed and used to transform E. coli cells harbouring a second expression plasmid containing the SS cDNA. Mutagenesis frequency was determined by quantifying the total number of bacterial colonies that did not significantly stain with I_2 (■) to the total number of colonies obtained (○).

were classified into six groups [20]. Groups I, II and III lacked significant levels of glycogen as determined by I_2 staining, but contained normal, intermediate and low AGP enzyme activities. Groups IV, V and VI stained very lightly owing to the low accumulation of glycogen, and their enzyme activities were normal, intermediate and low, respectively [20]. Similar mutagenesis and screening experiments have been conducted with the AGP SS cDNA expression plasmid, which resulted in the isolation of six mutant classes comparable with the ones generated with the LS cDNA plasmid [20a].

Analysis of LS and SS regulatory mutants

Of the various mutants obtained, the most interesting ones were those that belonged to groups I and IV. Mutants of these two groups exhibited wild-type levels of enzyme activity even though they were defective in glycogen accumulation. To determine the biochemical basis for the mutant phenotypes, AGP enzyme activities were partially purified by ammonium sulphate precipitation, heat treatment and by chromatography on a C-3 hydrophobic interaction column and/or ion-exchange column [21]. The enzymes from the various mutant lines were purified from 10 to 50% apparent homogeneity. Before assessing their kinetic properties, each enzyme sample was analysed by an immunoblot technique, using antibodies monospecific for the LS or SS for the presence or absence of proteolysis. If the samples indicated any evidence of proteolysis, they were discarded as ample evidence had been obtained that subtle alterations in enzyme structure can change the kinetic and allosteric regulatory properties of the enzyme (unpublished work of the laboratory of T.W. Okita, [22,23]).

Two LS mutant lines, M345 and M27, have been extensively characterized [20,24]. M345 (group I) and M27 (group IV) display substrate binding constants that do not differ substantially from the recombinant wild-type enzyme (Table 1). The K_M values for ATP, Glc 1-P and Mg^{2+} are either similar (M345) or somewhat lower (M27) than the wild-type enzyme. These results indicate that the inability of these mutant lines to accumulate significant amounts of glycogen is not due to a major change in the affinity of the enzyme for its substrates.

Further kinetic analysis revealed that the two LS mutant lines are defective in allosteric regulatory properties. A 3-PGA concentration of 0.6 mM ($A_{0.5}$) is required to activate the M27 enzyme to 50% of maximal activity. These $A_{0.5}$ values are approximately 6- to 10-fold higher than the value reported for the wild-type enzyme [18]. M345 needs even higher levels of 3-PGA for enzyme activity and has a $A_{0.5}$ for 3-PGA of 4.5 mM. Thus, the light-staining phenotype of M27 and the null-staining phenotype of M345 are most likely owing to the reduced allosteric activation of the enzyme by 3-PGA and, in turn, lower catalytic activity. In addition to 3-PGA activation, we evaluated the response of these mutant enzymes to Fruc 1,6-P_2 and phosphoenolpyruvate (PEP), which partially activate the wild-type enzyme 3- to 4-fold [18,25]. The M27 enzyme is totally insensitive to Fruc 1,6-P_2 while displaying a normal activation response to PEP. M345 shows no differences in Fruc 1,6-P_2 and PEP activation from the wild-type enzyme.

The activation of the wild-type enzyme by 3-PGA is reversed by P_i. As the inhibitory effects of P_i are more easily detected in the presence of 3-PGA, the enzymes were measured at a 3-PGA concentration corresponding to their measured $A_{0.5}$. At a concentration of 0.6 mM 3-PGA, the M27 enzyme is inhibited 50% ($I_{0.5}$) at a P_i concentration of 0.5 mM (Table 2, Figure 2). Likewise, the M345 enzyme displays an $I_{0.5}$ of 2.5 mM P_i when assayed in the presence of 4.4 mM 3-PGA. For comparison, the wild-type enzyme has a more pronounced resistance toward inhibition by P_i. In the presence of 0.1 mM 3-PGA, the wild-type enzyme has an $I_{0.5}$ for P_i of 0.7 mM. At 1.0 mM 3-PGA, the wild-type enzyme never attains 50% inhibition, even up to 5 mM P_i. Thus the mutant enzymes not only require higher levels of 3-PGA for activation but also have increased sensitivity to P_i inhibition.

Table 1

	ATP (mM)	Glc 1-P (mM)	Mg^{2+} (mM)
Wild type*	0.2–0.3	0.1–0.25	2.0–3.0
M345	0.35	0.3	2.6
M27	0.1	0.1	1.0
M947	0.3	0.2	3.5
D252N	0.2	1.2	4.5
A106T	1.4	0.25	6.4
D121N	0.5	2.0	6.4

**The wild-type kinetic values differ slightly among various investigators.*

Kinetic properties of AGP wild-type and mutant enzyme

Table 2

	$A_{0.5}$ 3 PGA (mM)	$I_{0.5}$ P_i (mM)
Wild type	0.1	0.7
M345	4.5	2.5
M27	0.6	0.5
M947	1.3	0.8

Allosteric properties of AGP wild-type and mutant enzymes

The molecular basis for the M27 and M345 biochemical phenotypes is single point mutations that changed single amino acid residues in each of the LSs of these AGPs. M27 has an Ala at position 413 in place of an Asp. The Asp is located 48 amino acids from the C-terminus and is part of a peptide that is highly conserved in every known LS sequence [17]. This conservation is consistent with a role for this peptide region in the normal allosteric functioning of AGP, as first shown by the activator-analogue studies on the spinach leaf enzyme [26]. Notably, three Lys residues on the LS and one Lys residue on the SS were modified by pyridoxyl 5-phosphate under reductive conditions [5,26]. All four Lys residues were protected by 3-PGA, but only the Lys residue of activator site 2 on the LS and the Lys residue of the SS were protected by P_i in competitive pyridoxal 5-phosphate labelling assays [5,26]. Asp-413 is immediately adjacent to Lys-414 of activator site 2 on the potato LS. The altered allosteric properties of M27 provide direct biochemical evidence that the C-terminal region of the LS is essential for 3-PGA activation of the AGP enzyme. Thus, two independent studies using

Figure 2

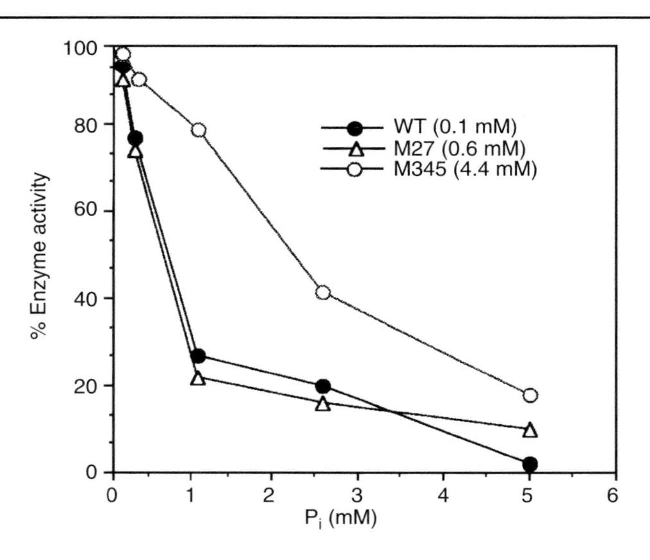

Inhibition of wild-type and mutant AGPs by P_i

Partially purified wild-type and mutant AGPs were assayed in the presence of increasing amounts of P_i, and the extent of ADPglucose formation measured. The enzymes were measured in the presence of 3-PGA at levels that yielded 50% of maximal catalytic activity ($A_{0.5}$). The 3-PGA concentration used for each enzyme is in parentheses; WT = wild type.

different techniques have identified this region as important in the interaction of AGP with its allosteric regulators.

M345 also contains a single amino acid replacement that is responsible for its altered allosteric regulatory properties. Unlike the structural change at the C-terminal end of M27, the mutation of M345 is located near the N-terminus of the LS where Pro-52 is replaced by Leu. Pro-52 is part of the sequence motif, Pro-Ala-Val (PAV), that is conserved in not only every LS sequence known to date but also in every SS sequence as well. The importance of this PAV motif in the SS is demonstrated by the kinetic properties of M947 [20a]. This enzyme contains a Ser in place of the equivalent Pro-52 of the LS. Similar to the M345 enzyme, the M947 enzyme possesses normal catalytic properties towards its substrates (Table 1) but is defective in allosteric activation by 3-PGA (Table 2). M947 requires 1.3 mM 3-PGA, a 13-fold increase over the wild-type enzyme.

The importance of the PAV sequence motif in allosteric regulation is also supported by studies with the *E. coli* enzyme. Replacement of the Ala of the PAV motif with a Thr results in a defect in allosteric regulation of the bacterial enzyme [27]. Moreover, at a position four residues from the N-terminal side of the PAV sequence motif is Lys-39, which is located in the activator binding site in the *E. coli* enzyme. The plant subunits display a conservation of charge at the position equivalent to Lys-39, suggesting that these residues may play a role in allosteric regulation. The regions flanking the PAV sequence are also highly conserved and contain a high frequency of positively charged residues that may be important in binding the plant and bacterial activators, 3-PGA and Fruc 1,6-P_2, respectively. The PAV motif may be an important structural motif in orienting residues in the correct conformation for efficient binding of these effectors. Irrespective of the exact role of the PAV sequence, these results indicate that the allosteric binding domain is at least partially conserved between the bacterial and plant enzymes. In addition, the plant's allosteric binding domain involves peptides located at both the N- and C-terminal ends of the LSs and SSs.

Analysis of substrate mutants

Examination of the group I and IV mutants obtained by chemical mutagenesis of the SS identified four distinct classes [20a]. In addition to the allosteric regulatory behaviour displayed by M947, three other classes were isolated that displayed altered kinetic constants for their substrates. Of the seven glycogen-deficient mutants isolated with wild-type antigen levels and high levels of enzymic activity, multiple mutants of each substitution were isolated, indicating that saturation mutagenesis of the SS cDNA was obtained. Specifically, two independent lines contained the replacement of Ala-106 with Thr (A106T), three independent lines contained the replacement of Asp-121 with Asn (D121N), while two independent lines contained the replacement of Asp-252 with Asn (D252N).

Kinetic analysis of representative members of each of these three mutant classes indicates defects primarily in binding of one or more substrates. The D252N enzyme exhibits a K_M of 1.2 mM for Glc 1-P, a value about 5- to 12-fold higher than the wild type enzyme. The binding constant for the other substrate, ATP, and the $A_{0.5}$ for 3-PGA do not differ from the wild-type enzyme. In addition, the D252N enzyme does show a doubling in the K_M for the cofactor Mg^{2+} from the wild type of 2.1 to 4.5 mM.

Enzyme containing the D121N substitution is also defective in binding of Glc 1-P, possessing a K_M some 8- to 20-fold higher than that of the wild-type enzyme. This mutant class also shows about a 2.5-fold increase in the binding constants for ATP and Mg^{2+} while the affinity constant for 3-PGA is not significantly different from the wild-type enzyme.

The A106T substitution results in several changes in both the catalytic and allosteric properties of the enzyme. The enzyme has an approx. 6.5-fold lower affinity for ATP when complexed with the cofactor Mg^{2+} than the wild type and a 3-fold lower affinity for K^+ATP. This substituted enzyme also has a lower affinity than the wild-type towards the cofactor Mg^{2+} and the allosteric activator 3-PGA. However, this substitution has no effect on the binding constant for Glc 1-P ($K_M = 0.25$).

The biochemical defects exhibited by these three substrate mutant classes are consistent with the available information on the structure–function relationships of AGP. Tyr-114 has been identified to be located at or near the ATP binding site in the *E. coli* AGP [16]. Replacement of Tyr-114 with Phe in the *E. coli* enzyme resulted in the formation of an enzyme that not only has lower affinity for ATP but also for the allosteric activator Fruc 1,6-P_2 and the cofactor Mg^{2+} [28]. Two of the residue changes identified in this study, A106T and D121N, are within the conserved region of substrate binding equivalent to Tyr-114 of the *E. coli* enzyme. As first observed for the bacterial enzyme, these mutations not only result in enzymes with altered affinities for the substrate ATP, but also for Glc 1-P (D121N), and 3-PGA and Mg^{2+} (A106T). It is interesting that independently derived substitutions in bacteria and plants have identified this region around residue Tyr-114 as being involved in ATP binding.

Residues located at or near the Glc 1-P substrate binding sites have also been identified. The Lys-195 of the *E. coli* enzyme has been shown to be at or near the Glc 1-P binding site [29]. Site-directed mutagenesis of the SS Lys-188, equivalent to the *E. coli* Lys-195, to Glu increased the K_M for Glc 1-P from 80 μM to over 45 mM [5]. Located near Lys-188 is Asp-252. Replacement of Asp-252 with Asn produces an enzyme with a lower affinity for Glc 1-P without affecting the interaction with any other substrate or allosteric effector. Since the enzyme encoded by the D252N substitution is solely affected in its affinity for Glc 1-P, Asp-252 may be near an uncharacterized substrate binding site or be required for conformation of the Glc 1-P binding domain.

Role of LS and SS in enzyme function

Based on the types of mutations generated and identified by this biochemical genetic approach, the roles of the LS and SS are beginning to emerge. Two of the LS mutants analysed are defective only in allosteric regulatory behaviour. Preliminary characterization of two other group IV LS mutants indicate defects in allosteric regulation as well. The isolation of M345 and M27 provides direct evidence that the LS is involved in the allosteric properties of the heterotetrameric holoenzyme. In contrast, almost all of the group I and IV SS mutant lines examined are defective mainly in catalysis. The one exception is M947, which is the SS mutation analogous to the LS mutant M345. M345, however, has a much greater defect in 3-PGA activation than M947. Overall, these results support the view that the LS and SS may not have identical roles in enzyme function. This hypothesis is supported by the capacity of the SS, when expressed in the absence of the LS, to form an active homotetrameric enzyme. The SS homotetramer, however, requires very high levels of 3-PGA to achieve activation [30]. In contrast, the LS alone is unable to assemble into a catalytically active enzyme. These observations, and the results of our mutant analysis, suggest that the two subunit types possess distinct roles in the functioning of this enzyme. The SS may play a more dominant role in catalysis while the LS may serve to increase the sensitivity of the SS towards activation by 3-PGA.

Molecular evolution of the SS enzyme

The ability of the SS to assemble into an active enzyme, albeit with reduced sensitivity to 3-PGA activation, raises questions on the origin and evolution of the plant AGP subunits, especially the SS. To determine the type of changes that may have occurred during the evolution that resulted in the formation of a catalytic SS dependent on the LS for allosteric properties, we initiated a mutagenesis programme on the SS sequences alone. Expression of the wild-type SS sequences results in a null-staining phenotype when cells are exposed to I_2. When mutated by hydroxylamine, several cell lines displayed varying levels of I_2 staining. One line, TG15, was selected for further study. Analysis of the regulatory properties of the TG15 enzyme indicated an increased sensitivity towards activation by 3-PGA and other organic phosphate esters. The TG15 enzyme has an $A_{0.5}$ for 3-PGA of 0.1 mM, a value equal to the recombinant wild type SS enzyme. In addition, it shows enhanced activation with other glycolytic intermediates. Fruc 6-P at 5 mM was able to activate the enzyme 16.2-fold whereas the wild-type SS enzyme was activated only about 1.5-fold. Likewise, PEP was able to activate the TG15 enzyme at more than double the level exhibited by the wild-type SS enzyme. The increased sensitivity to activation by 3-PGA and Fruc 6-P was due to two point mutations where Leu-48 is replaced by Phe and Val-59 is replaced by Ile. Leu-48 is only a single residue removed from the PAV sequence motif in the SS and supports the role of this peptide region in allosteric regulation. It is interesting to note that a Phe residue is located at the equivalent position of Leu-48 in the bacterial enzyme.

Concluding remarks

The capacity of the plant AGP subunits to be expressed and assembled into a heterotetrameric enzyme capable of complementing a mutation in the *E. coli glg*C gene has provided a novel and powerful means to assess the structure–functional aspects of AGP, heretofore not possible by existing approaches. Based on the analysis of mutants that have lost the ability to complement *glg*C, valuable information on the possible roles of these subunits in enzyme function and the identification of residues involved in catalysis and allosteric regulation has been obtained. Using these tools additional insights are expected to be obtained on structural aspects, e.g. regions of the LS and SS that interact with one another.

Moreover, in addition to the down-regulatory mutant lines such as M27 and M345 obtained, new strategies can be devised for the isolation of up-regulatory AGP enzymes that require very little or no activator for maximal catalytic activity. The isolation of these up-regulatory lines will be important in addressing critical questions on carbon partitioning between sucrose and starch and in enhancing starch production in developing sink organs.

The use of *E. coli* to serve as a host for genetic studies will also have utility in the study of the other starch biosynthetic enzymes. *E. coli* mutants in *glg*A (glycogen synthase) and *glg*B (branching enzyme) are available and can be used in a strategy similar to that discussed here to identify mutations in starch synthases and branching enzymes. In addition to the generation of new enzyme variants that may lead to new types of starches, the system will be invaluable in assessing the interaction of specific starch synthases and branching enzymes and in identifying their specific roles in the formation of the starch granule.

The research on the structure–function of AGP is supported by grants from the Department of Energy Grant DE-FG0687ER136 and NASTI, RDA, Korea, and CAHE Project 0119. M.J.L., P.S. and R.W. are recipients of a National Institutes of Health Traineeship in Protein Biotechnology.

References

1. Nelson, O. and Pan, D. (1995) Annu. Rev. Plant Physiol. Plant Mol. Biol. **46**, 475–496
2. Nelson, O.E. (1982) in Advances in Cereal Science and Technology (Pomeranz, Y., ed.), pp. 41–71, American Association of Cereal Chemists, St. Paul
3. Libessart, N., Maddelein, M.-L., Van den Koornhuyse, N., Decq, A., Delrue, B., Mouille, G., D'Hulst, C. and Ball, S. (1995) Plant Cell **7**, 1117–1127
4. Wurzburg, O.B. (1986) Modified Starches: Properties and Uses, pp. 277, CRC Press, Boca Raton, FL
5. Preiss, J. and Sivak, M. (1995) in Photoassimilate Distribution in Plants and Crops: source–sink relationships (Zamski, E. and Schaffer, A.A., eds.), pp. 139–168, Dekker, New York
6. Martin, C. and Smith, A.M. (1995) Plant Cell **7**, 971–985
7. Larsson, C.-T., Hofvander, P., Khoshnoodi, J., Ek, B., Rask, L. and Larsson, H. (1996) Plant Sci. **117**, 9–16
8. Preiss, J. (1992) in Oxford Surveys of Plant Molecular and Cellular Biology, (Miflin, B.J., ed.), Vol. 7, pp. 59–114, Oxford University Press, Oxford
9. Preiss, J., Hutney, J., Smith-White, B., Li, L. and Okita, T.W. (1991) J. Pure Appl. Chem. **63**, 535–544
10. Preiss, J., Ball, K., Smith-White, B., Iglesias, A., Kakefuda, G. and Li, L. (1991) Biochem. Soc. Trans. **19**, 539–547
11. Guan, H.P. and Preiss, J. (1993) Plant Physiol. **102**, 1269–1273
12. Takeda, Y., Guan, H.P. and Preiss, J. (1993) Carbohydr. Res. **240**, 253–263
13. Sumner, J.B. and Somers, G.F. (1944) Arch. Biochem. **4**, 4–7
14. Pan, D. and Nelson, O.E. (1984) Plant Physiol. **74**, 324–328
15. James, M.G., Robertson, D.S. and Myers, A.M. (1995) Plant Cell **7**, 417–429
16. Preiss, J. and Romeo, T. (1994) in Progress in Nucleic Acid Research and Molecular Biology, (Cohn, W.E. and Moldave, K., eds.), Vol. 47, pp. 299–329, Academic Press, San Diego, CA
17. Smith-White, B.J. and Preiss, J. (1992) J. Mol. Evol. **34**, 449–464
18. Iglesias, A.A., Barry, G.F., Meyer, C., Bloksberg, L., Nakata, P.A., Greene, T., Laughlin, M.J., Okita, T.W., Kishore, G.M. and Preiss, J. (1993) J. Biol. Chem. **268**, 1081–1086
19. Suzuki, D.T., Griffin, A.J.F., Miller, J.H. and Lewontin, R.C. (1989) in Introduction to Genetic Analysis, pp. 475–499, Freeman, New York
20. Greene, T.W., Chantler, S.E., Kahn, M.L., Barry, G.F., Preiss, J. and Okita, T.W. (1996) Proc. Natl. Acad. Sci. U.S.A. **93**, 1509–1513

20a. Laughin, M.J., Payne, J.W. and Okita, T.W. (1998) Phytochemistry **47**, 621–629

21. Okita, T.W., Nakata, P.A., Anderson, J.M., Sowokinos, J., Morell, M. and Preiss, J. (1990) Plant Physiol. **93**, 785–790
22. Plaxton, W.C. and Preiss, J. (1987) Plant Physiol. **83**, 105–112
23. Kleczkowski, L.A., Villand, P., Luthi, E., Olsen, O.-A. and Preiss, J. (1993) Plant Physiol. **101**, 179–186
24. Greene, T.W., Woodbury, R.L. and Okita, T.W. (1996) Plant Physiol. **112**, 1315–1320
25. Sowokinos, J.R. and Preiss, J. (1982) Plant Physiol. **69**, 1459–1466
26. Ball, K. and Preiss, J. (1994) J. Biol. Chem. **269**, 24706–24711
27. Meyer, C.R., Ghosh, P., Nadler, S. and Preiss, J. (1993) Arch. Biochem. Biophys. **302**, 64–71
28. Kumar, A., Ghosh, P., Lee, Y., Hill, M.A. and Preiss, J. (1989) J. Biol. Chem. **264**, 10464–10471
29. Hill, M.A., Kaufmann, K., Otero, J. and Preiss, J. (1991) J. Biol. Chem. **266**, 12455–12460
30. Ballicora, M.A., Laughlin, M.J., Fu, Y., Okita, T.W., Barry, G.F. and Preiss, J. (1995) Plant Physiol. **109**, 245–251

Prospects of engineering heavy metal detoxification genes in plants

David W. Ow
Plant Gene Expression Center, USDA-Agricultural Research Service, 800 Buchanan St., Albany, CA 94710, U.S.A.

Abstract

Unlike compounds that can be broken down, the remediation of most heavy metals and radionucleotides requires removal from contaminated sources. Plants can extract inorganics, but effective phytoextraction requires plants that produce high biomass, grow rapidly and possess high uptake capacity for the inorganic substrate. The existing hyperaccumulator plants must be bred for either increased growth and biomass, or the hyperaccumulation traits must be engineered into fast growing, high-biomass plants. The latter approach requires fundamental knowledge of the molecular mechanisms in the uptake and storage of inorganics. Much has been learned in recent years of how plants and certain fungi chelate and transport cadmium. This progress has been facilitated by the use of the fission yeast *Schizosaccharomyces pombe* as a model system. As target genes are identified in a model organism, their sequences provide opportunities for engineered expression in a heterologous host and aid the search for homologous genes in complex organisms.

Introduction

Heavy metals released into the environment can have long-term effects on human health. Unlike organic compounds that can be mineralized, the remediation of inorganics requires removal or conversion into a biologically inert form. Some elements, such as Se and Hg, can be removed through biovolatilization [1–3]. Most other metals, however, require the physical extraction from soil and water systems. The cost of conventional engineering methods for the remediation of soil contamination starts at US$250 000 acre^{-1}. Owing to this prohibitive cost, there is increased support to develop new remediation strategies, including the concept that genetically modified plants can be a cost-effective agent for extracting heavy metals from contaminated sources [4].

A wide variety of bacterial, fungal, algal and plant systems are capable of concentrating toxic metals from their surroundings. It is difficult to retrieve small organisms from soil. In contrast, contaminant-laden plants can be readily harvested, especially if the metals are translocated to aerial parts. The contaminated biomass can then be disposed of after a substantial reduction in volume and weight. For some metals (e.g. Ni, Zn, and Cu) the value of the reclaimed material provides additional incentive for metal recycling. The rate of metal removal depends on the amount of harvested biomass, the number of harvests per year and the metal concentration in the harvested material. Decontaminating a site in a reasonable number of years requires plants with high biomass and high metal accumulation. Plants that exhibit the latter trait do exist and are known as metal-hyperaccumulators [5]. However, they are rather slow growing plants with low biomass and restricted element selectivity. Moreover, there is little known about their agronomics, genetics, plant breeding potential and disease spectrum.

The harvestable aerial tissues of known hyperaccumulators can concentrate metals ranging from 0.2% (e.g. Cd) to 5% (e.g. Zn, Ni, Mn) of their dry weight [6], which is generally two or more orders of magnitude greater than the concentrations found in the more familiar crop plants. This comparison assumes that both plant types are capable of growing in the metal contaminated soil, which is not necessarily the case for most plants that do not tolerate high metal content. Despite this impressive difference in metal uptake, the annual biomass yield of hyperaccumulators is generally one to two orders of magnitude lower than those of robust crop plants. Thus, if the high yield and hyperaccumulation traits could be bred into a single plant, the effectiveness of metal extraction could be increased by one to two orders of magnitude. Plant-based remediation efforts would have a much more practical time-frame, of the order of a decade rather than the current estimates of a century or more.

Genetic potential for novel metal hyperaccumulation traits

Plant breeding is a conventional approach to combining desirable traits, in this case metal hyperaccumulation and high yield. However, genetic improvements are limited to the traits that nature provides within sexually compatible species. Modern molecular biology, however, can overcome the sexual barrier through direct gene transfer. In principle, the genetic engineering of novel hyperaccumulation traits directly into fast growing, high-biomass plants should be within the realm of possibility. However, this will require fundamental knowledge of the molecular mechanisms that plants use to accumulate toxic ions. Target areas for research include: (1) metal bioavailability through plant growth, (2) metal uptake by plant roots, (3) metal translocation to aerial harvestable tissue and (4) metal tolerance and storage at the cellular level. Research in the first three areas is expected to enhance plant metal uptake. However, without enhancing cellular tolerance mechanisms, hyperaccumulation of metals would merely speed up cellular poisoning of aerial photosynthetic tissue. This effect would, of course, limit biomass and growth rate, the very factors crucial for high capacity metal extraction. Thus, research into the cellular mechanisms of metal tolerance is crucial for metal phytoremediation efforts. In this light, it is important to consider appropriate model systems to expedite basic knowledge in this area.

Model organisms for defining cellular tolerance mechanisms

Much of our recent knowledge of the molecular and cellular mechanisms of cadmium chelation and storage in plants stems from studies of the fission yeast *S. pombe* [7]. Certain fungi share with plants the same cellular response to metal stress. The relative ease with which fission yeast can be manipulated genetically permits a more rapid characterization of the molecular processes. Once target genes are cloned from a model organism, their sequences can be modified for gene transfer to a heterologous host, or they can be used to search for homologous genes in complex organisms. Moreover, as plant nutrient uptake is intrinsically linked to associated rhizospheric fungi, elucidating fungal metal metabolism affords additional opportunities for engineering rhizospheric micro-organisms to assist in plant metal extraction.

Metal-induced chelators

Intracellular chelation is a well-described mechanism in metal tolerance and a key step for metal extraction. In response to heavy metals, animals and certain fungi synthesize small cysteine-rich proteins known as metallothioneins [8]. There have been numerous reports of engineering the production of animal metallothioneins in plants [9–13]. Varying degrees of increased metal tolerance have been achieved, but none reported substantial increases in overall metal uptake. This may be because plants do not naturally use metallothioneins for metal detoxification. Therefore, the anomalous production of unfamiliar ligands does not lead to further transport and storage of the bound metals. Although metallothionein genes have also been described in plants [14], the proteins have not yet been found, casting doubt on their role in metal tolerance. Instead, heavy metals in plants and some fungi induce the production of metal-binding peptides commonly known as phytochelatins [15,16]. *S. pombe* is one fungus that shares this response with higher plants.

Derived from glutathione, phytochelatins (PCns) have the general structure of $(\gamma\text{-Glu-Cys})_n\text{Gly}$, where n is generally from 2 to 5, but as many as 11 γ-Glu-Cys units have been described [17]. Some PCn-related peptides lack the carboxyl-terminal Gly or have instead β-Ala, Ser or Glu. However, these variant peptides are usually found in lower abundance compared with the PCns. As with metallothioneins, the cysteine residues of PCns form thiolate bonds with the metal cations. A large variety of metals induce the synthesis of PCns, but formation of a PCn–metal complex has largely been examined with Cd^{2+} and Cu^{2+}. Several reports show that PCns also form complexes with Ag^{+}, Hg^{2+}, Pb^{2+} and Zn^{2+} [18–21]. In the case of Cd^{2+}, two PCn–metal complexes can be isolated from cells exposed to Cd^{2+}: a low relative molecular mass (low-M_r) PCn–Cd complex and a more stable PCn–CdS complex with a high relative molecular mass (high-M_r) that contains acid-labile sulphide (S^{2-}). The appearance and the location of the two complexes suggest that the PCn–Cd complex acts as a cytoplasmic scavenger and carrier of metals to the vacuole, where they are stored as stable sulphide-rich chelates.

We have isolated a number of *S. pombe* mutants that are impaired in metal detoxification. These mutants are hypersensitive to Cd^{2+} and fail to form wild-type levels of one or both of the PCn-bound Cd complexes. The sections that follow summarize our current understanding of the cadmium detoxification process.

Vacuolar storage

Through the analysis of LK100, a mutant that fails to form the high-M_r PCn–CdS complex, we isolated a gene encoding a vacuolar membrane protein that is a member of the ATP-binding cassette (ABC)-type transporter family [22]. This gene was named *hmt1* (for heavy metal tolerance). We found that the HMT1 protein transports cytoplasmic PCn–Cd complex into the vacuole [23]. Table 1 summarizes the ATP-dependent transport activities of vacuolar vesicles from HMT1-proficient and -deficient strains. ATP-dependent uptake of ^{35}S-labelled PCn–Cd was seen with vesicles from the HMT1$^+$, but not the HMT$^-$ strain. Only low activity was observed with PCn–CdS as a substrate, most likely representing the transport of PCn–Cd derived from spontaneous disassociation of the high-M_r complex. Apo-PCn peptides were efficiently transported, but this activity could represent the prior complexation of apo-peptides with traces of metals in the vacuolar vesicle preparation. A high Mg^{2+} concentration was provided for the

Table 1

Substrate	Inhibitor	HMT1$^+$	HMT1$^-$
[^{35}S]PCn–CdS	None	–	–
[^{35}S]PCn–Cd	None	+	–
	Bafilomycin	+	–
	Nigericin	+	–
	Vanadate	–	–
	Antibody	–	–
[^{35}S]Apo-PCn	None	+?	–
[^{109}Cd]PCn–Cd	None	+	+
	Bafilomycin	+	–
	Bafilomycin and vanadate	–	–
$^{109}Cd^{2+}$	None	+	+
	Bafilomycin	–	–
	Nigericin	–	–
	Vanadate	+	+

ATP-dependent uptake by vacuolar vesicles

Vacuolar vesicles were prepared from HMT$^+$ or HMT$^-$ strains. Presence (+) or absence (–) of ATP-dependent transport activity was determined by uptake of radiolabelled substrates in the presence or absence of indicated inhibitors. Adapted from Ortiz and co-workers [23]. The ? indicates uncertainty with transport of apo-PCn peptides as complex formation with trace metals has not been ruled out.

ATP-dependent reaction. Transport of PCn–Cd was independent of the pH gradient as it was unaffected by the vacuolar ATPase inhibitor bafilomycin or the H^+–K^+ ionophore nigericin. It was reduced significantly, however, by antibodies directed against HMT1, and by vanadate, an inhibitor known to affect many ABC-type transporters.

When ^{109}Cd-labelled PCn–Cd was provided as the substrate, uptake of the label was independent of HMT1. Further examination showed that uptake of the $^{109}Cd^{2+}$ cation is itself independent of HMT1 and vanadate, but is abolished by the inhibitors bafilomycin and nigericin, which dissipate the pH gradient. This suggests the presence of an antiport that can use Cd^{2+} as a substrate. Hence, the activity observed with ^{109}Cd-labelled PCn–Cd was probably due to free Cd^{2+} disassociated from the complex. In the presence of bafilomycin, which blocked the antiport activity, uptake of the label from ^{109}Cd-labelled PCn–Cd became dependent on HMT1, and was significantly reduced by the presence of both bafilomycin and vanadate. The data are consistent with the interpretation that HMT1 co-transports PCn peptides and Cd^{2+} as a complex.

Higher plants have also been found to transport PCn–Cd to the vacuole through what appears to be an ABC-type transporter [24], but the plant transporter gene has not yet been cloned. When the HMT1 protein was overproduced in *S. pombe*, the cells showed enhanced cadmium accumulation and tolerance, presumably due to more effective sequestering of the metal. Taken together, these findings suggest that hyperproduction of a PCn transporter in the plant vacuolar membrane might similarly enhance the vacuolar sequestration of toxic metals, and thereby enhance greater metal-extraction from contaminated sources [7]. The activity of a Cd^{2+} antiport has also been reported in plants [25]. As the high-M_r complex incorporates additional Cd^{2+} ions, it is believed that the additional ions are transported into the vacuole through this route (Figure 1).

Cadmium-inducible sulphide pathway

The stable high-M_r complex found in the vacuole has sulphide ions incorporated into the PCn–Cd complex. In the absence of sulphide, the Cys to Cd ratio has been estimated to be from 2 to 4 [26]. Upon sulphide addition, the Cys to Cd ratio drops to ~1. Thus, the incorporation of sulphide increases the Cd storage capacity per PCn peptide. Furthermore, the high-M_r complex, which appears to consist of CdS crystallites surrounded by PCn peptides [27], is more stable than the low-M_r complex in acidic environments [28], such as in the lumen of the vacuole. As formation of the PCn–CdS complex is crucial to metal storage, the source of the sulphide ion is an important issue to address. Under metal stress, cellular sulphide levels increase at least 7-fold when the cells are grown in rich medium (and higher when grown in minimal medium). Since plants and micro-organisms assimilate inorganic sulphur through a pathway in which sulphate (SO_4^{2-}) is incorporated into ATP and subsequently reduced to form sulphite (SO_3^{2-}) and then sulphide (Figure 1), it is possible that the increased sulphide production is due to increased activity of this assimilatory reduction pathway. However, the analysis of one

Figure 1

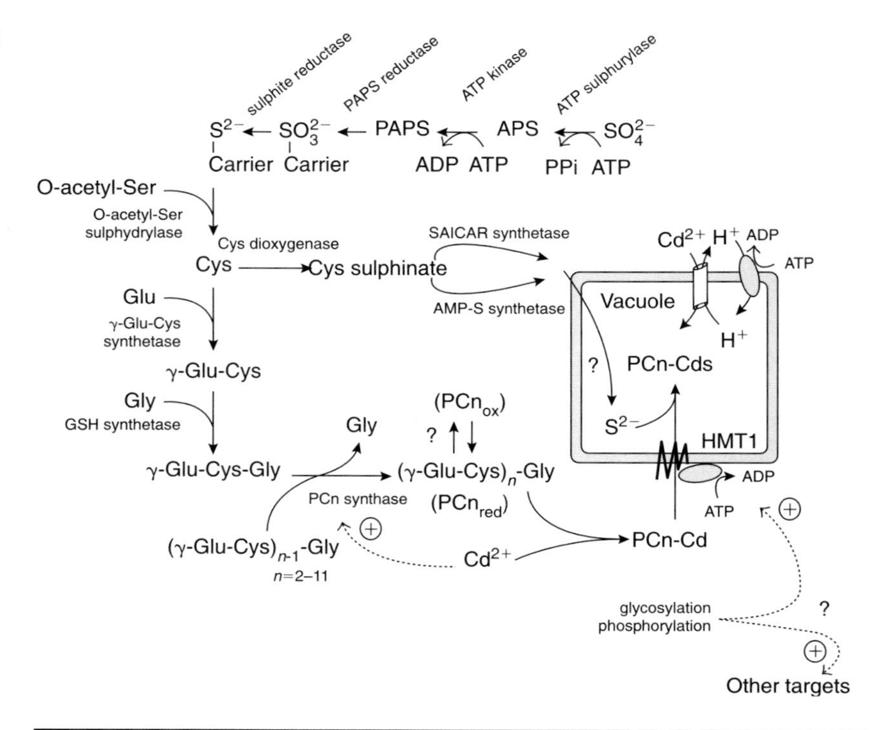

Molecular model of cadmium detoxification

Cd^{2+} activates the enzymic synthesis of PCn peptides. The peptides are kept from being oxidized by a hypothetical PCn reductase (suggested by mutant JS563). The reduced PCn peptides chelate Cd^{2+} and the absence of free Cd^{2+} terminates enzymic synthesis. The PCn–Cd complex, with a Cys:Cd^{2+} ratio of from 2:1 to 4:1, is transported into the vacuole by the HMT1 protein (shown by mutant LK100); S^{2-} and additional Cd^{2+} are added to form a more acid-stable high-M_r complex, with Cys:$Cd^{2+} \approx 1:1$. Most likely, the additional Cd^{2+} ions enter through a cation antiport, requiring the pH gradient generated by a vacuolar H^+-ATPase. The source of S^{2-}, which is Cd inducible, is not from the SO_4^{2-} assimilatory sulphate reduction pathway (suggested by mutant DS12), but through a novel pathway derived from cysteine, and requires enzymes that participate in purine biosynthesis (suggested by mutant LK69). The proposed route starts with formation of Cys sulphinate from Cys, followed by incorporation of Cys sulphinate into purine precursors by SAICAR synthetase and AMP-S synthetase. HMT1 and possibly other proteins involved in metal stress may be affected by glycosylation (suggested by mutant JS618) and phosphorylation (suggested by mutant JS237). Cadmium stress appears to involve a signal transduction pathway that utilizes cAMP and Ca^{2+} as secondary messengers. Abbreviations used: APS, adenosine 5′-phosphosulphate; PAPS, 3′-phosphoadenosine 5′-phosphosulphate; AMP-S and SAICAR, see Figure 2 legend.

particular mutant, LK69, revealed that Cd-inducible sulphide production requires genes in the purine biosynthetic pathway [29,30].

Figure 2 shows the relevant segment of the purine biosynthesis pathway (abbreviations defined in the legend). LK69, which is Cd-hypersensitive and deficient in accumulation of PCn–CdS complex, was found to harbour genetic lesions in both *ade2* (encoding AMP-S synthetase) and *ade6* (encoding AIR

Figure 2

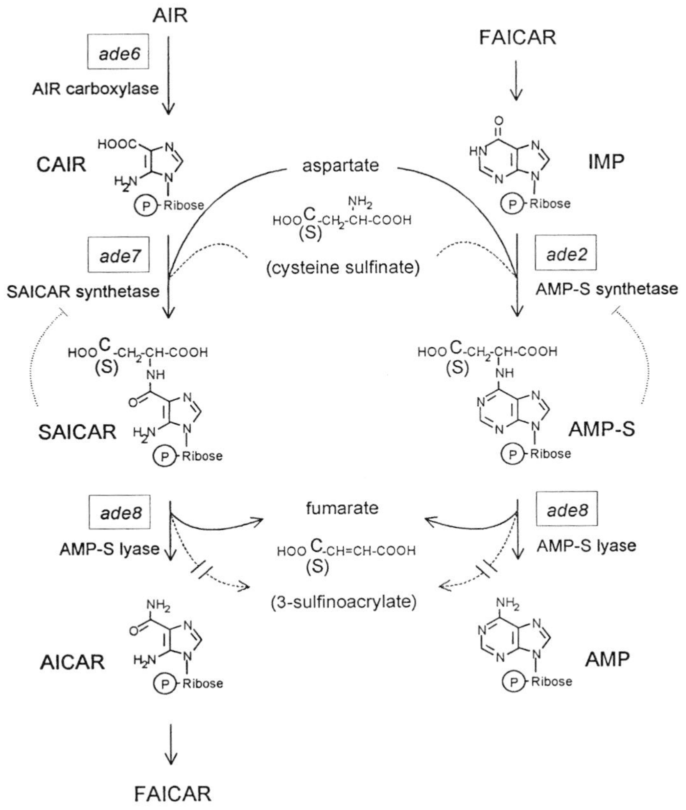

Relevant segment of the purine biosynthesis pathway and the proposed biosynthesis of cysteine sulphinate-derived products (dashed lines)

The pathway from AIR to AMP is shown together with hypothetical sulphur analogue products (in parentheses). Formation of S-derivatives by SAICAR synthetase and AMP-S synthetase have been shown in vitro. However, AMP-S lyase does not react with these S-derivatives. We propose that AMP-S lyase activity is needed to prevent product inhibition (dotted lines) of SAICAR synthetase and AMP-S synthetase, rather than for the formation of 3-sulphinoacrylate. Abbreviations used: AIR, aminoimidazole ribonucleotide; CAIR, carboxyaminoimidazole ribonucleotide; SAICAR, succinoaminoimidazolecarboxamide ribonucleotide; AICAR, aminoimidazolecarboxamide ribonucleotide; FAICAR, formamidoimidazole-carboxamide ribonucleotide; IMP, inosine monophosphate; AMP-S, adenylosuccinate. Adapted from Speiser and co-workers [29] and Juang and co-workers [30].

carboxylase). This led to the discovery that an *ade2*$^-$*ade7*$^-$ double mutant also exhibited the same phenotype as LK69. However, a single lesion in either *ade2*, *ade6* or *ade7* does not cause Cd hypersensitivity. Although AIR carboxylase acts at a step upstream of the AMP-S synthetase reaction, a mutation blocking IMP production via this linear pathway does not exert an epistasis effect on AMP-S synthetase because adenine can be converted into IMP via a salvage pathway. Hence, the IMP to AMP reactions are operational in an *ade6*$^-$ or *ade7*$^-$ mutant grown with adenine supplementation, just as the reactions leading to IMP production catalysed by AIR carboxylase are operational in an *ade2*⁻ background. If each segment of the pathway catalyses a reaction that can be complemented by the other, then blockage of both segments of the pathway would be needed to produce a deficiency phenotype, in this case Cd hypersensitivity.

The reactions performed by the two segments are indeed similar. The conversion of CAIR into SAICAR is analogous to the conversion of IMP into AMP-S in that both reactions incorporate aspartate on to a nucleotide substrate. Cysteine sulphinate, an oxidative product of cysteine, can replace aspartate in an *in vitro* reaction catalysed by the AMP-S synthetase from *Azotobacter vinelandii* [31]. If this analogue substrate is also utilized by SAICAR synthetase, then it may be possible that the sulphur-analogue intermediates from either segment of the pathway serve an essential role for Cd tolerance, e.g. as sulphide carriers or donors. Additionally, AMP-S lyase might catalyse the release of the S-analogue of fumarate, 3-sulphinoacrylate, as an intermediate in the formation of sulphide for the PCn–CdS complex. An *ade8*⁻ mutant is indeed sensitive to Cd, but exhibits only a slower rate PCn–CdS complex formation. To account for the *ade8*⁻ phenotype, one possibility would be that AMP-S lyase performs a function essential to the biogenesis of the PCn–CdS complex, and the slow rate of high-M_r complex formation could be due to leaky synthesis of active enzyme. An alternative explanation would be that a lack of AMP-S lyase leads to hyperaccumulation of AMP-S and SAICAR, which could feedback inhibit AMP-S synthetase and SAICAR synthetase, respectively. In this latter scenario, a lesion in *ade8* would mimic double lesions of *ade2* and *ade7*. Assuming that product inhibition of enzyme function is less effective than a genetic blockage of enzyme synthesis, this could account for the 'leaky' accumulation of the high-M_r complex seen in the AMP-S lyase-deficient mutant.

In testing the above hypothesis, we examined the activities of AMP-S synthetase, SAICAR synthetase and AMP-S lyase. The reaction of AMP-S synthetase with [^{35}S]cysteine sulphinate generated a novel ^{35}S-labelled compound detected by thin layer chromatography. Subsequent addition of AMP-S lyase or crude extracts to the reactions failed to produce evidence for the formation of [^{35}S]sulphinoacrylate. Similarly, SAICAR synthetase also formed a novel radiolabelled compound from [^{35}S]cysteine sulphinate that failed to react further upon addition of AMP-S lyase or crude extracts. Thus, both genetic and biochemical data are consistent with a model that AMP-S synthetase and SAICAR synthetase can incorporate a sulphur analogue of aspartate on to purine intermediates. Sulphur addition to a purine molecule is not without precedent, as assimilation of sulphate begins through incorporation into ATP to form adenosine 5′-phosphosulphate (APS, Figure 1).

Assimilatory sulphate reduction pathway required for Pb-induced, but not Cd-induced sulphide production

The above proposal of a novel Cd-inducible pathway for sulphide production is consistent with the behaviour of a mutant blocked in assimilatory sulphate reduction. This mutant, DS12, harbours a lesion in a gene with homology to sulphite reductase genes (Figure 1). An engineered disruption of this gene resulted in Cys auxotrophy, as expected for the inability to convert sulphite into sulphide. As with all Cys autotrophs, the gene disruption strain DS35 is hypersensitive to Cd and fails to produce PCns despite Cys supplementation. Presumably, the slow rate of Cys transport cannot provide the high Cys level required for PCn production. Upon Cys supplementation, however, the mutant can nonetheless respond to Cd-induced production of sulphide (Figure 3). This supports the view that sulphide induction during Cd stress proceeds through a pathway starting with Cys as the source. Sulphide production can also be induced by Pb, leading to PbS precipitates. However, this Pb-induced sulphide production is abolished in DS35 (Figure 3). It thus appears that while the assimilatory sulphate reduction pathway is not needed for Cd-induced sulphide production, it is nonetheless required in the case of Pb.

PCn accumulation requires a FAD/NAD-linked disulphide reductase

JS563 was isolated as a mutant defective in the production of Cd-bound PCn complexes. Interestingly, this Cd-hypersensitive mutant also hyperproduces sulphide. The gene responsible for this phenotype was cloned by genetic complementation, which restored JS563 to a wild-type level of Cd tolerance, normal accumulation of Cd-bound PCn complexes and normal production of sulphide. The encoded protein shares sequence similarity with members of the FAD/NAD-linked disulphide reductase family. These proteins catalyse the reduction of oxidized substrates, such as cytochrome or glutathione. There is sequence

Figure 3

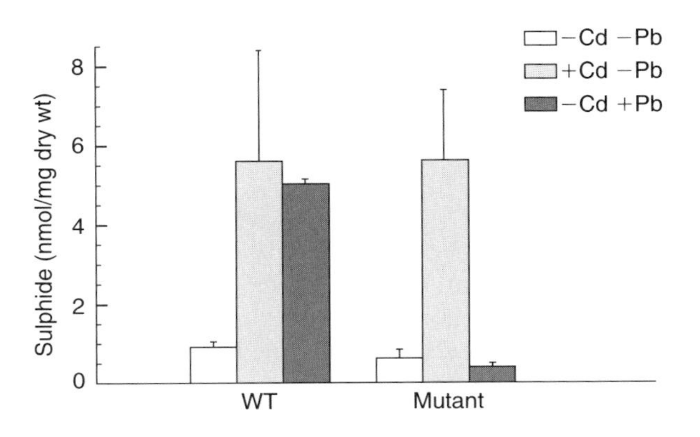

Direct need for assimilatory sulphate reduction for Pb-induced, but not Cd-induced sulphide production

A sulphite-reductase mutant, auxotrophic for cysteine, can produce Cd-induced levels of sulphide during Cd stress (+ Cd) when provided with cysteine. During Pb stress (+ Pb), however, the mutant is defective in Pb-induced sulphide production (WT = wild type).

similarity with glutathione reductase (29% similar). However, this protein is not likely to reduce glutathione because the mutant produces reduced glutathione and because there are some critical differences in the arrangement of the predicted functional sites (redox active cysteines) in the two proteins. Our current hypothesis is that it might be involved in the reduction of PCn peptides. If PCn peptides become oxidized, formation of intra- and inter-molecular disulphide bridges would prevent the sulphydryl groups from coordinating with Cd^{2+} (Figure 1). Recently, we have localized the protein to the mitochondria and it is possible that the electrons for reduction of PCn peptides originate from this organelle. In the mutant, the inability to form reduced PCn peptides would account for the lack of bound Cd complexes. As PCn peptide production is regulated by feedback control, i.e. free metal ions induce enzymic synthesis, the inability of oxidized PCn peptides to bind metals would cause further synthesis of PCn peptides. Degradation of large amounts of oxidized PCn peptides could account for the aberrant high sulphide level found in JS563.

PCn–CdS complex formation requires a protein similar to mannose-1-phosphate guanyltransferase

Recent work on JS618, a mutant unable to accumulate the high-M_r PCn–CdS complex, resulted in the isolation of a gene whose protein product has sequence similarity with mannose-1-phosphate guanyltransferase, an enzyme that converts mannose-1-phosphate into GDP-mannose. GDP-mannose is a substrate for glycosyltransferases in protein glycosylation and polysaccharide biosynthesis. Our current thinking is that the mutation abolishes the glycosylation of a critical protein needed for formation of the high-M_r complex. As many ABC-type transporters (e.g. the cystic fibrosis transmembrane conductance regulator and the multidrug resistance pump) are glycoproteins, we are investigating whether this mutation prevents HMT1 targeting or activity.

Signal transduction in metal accumulation

The analysis of mutant JS237 suggests that signal transduction involving cAMP and Ca^{2+} is needed for vacuolar accumulation of cadmium. This Cd hypersensitive mutant fails to accumulate the high-M_r complex. The complementing DNA revealed two genes encoding proteins with strong sequence similarities to two human proteins described in recent literature: BTF3 and WASP (Wiskott–Aldrich Syndrome protein). It is most unusual that each protein can apparently be encoded on a separately transcribed mRNA or the two proteins encoded together on one bicistronic mRNA.

BTF3 was initially thought to be a basal transcription factor because it co-purified with RNA polymerase II [32]. Recently, BTF3 was found to be the β subunit of a protein complex known as NAC, which binds nascent polypeptide chains as they are formed by the ribosome [33]. NAC competes with the signal recognition particle (SRP). If the nascent chain has a signal sequence, then SRP binds to that signal and the polypeptide is directed to the endoplasmic reticulum (ER). However, if NAC is not present, then the SRP will bind any polypeptide, even those without the signal sequence, and direct it to the ER. A mutation in NAC would be expected to result in missorting of proteins. In humans, there are at least nine BTF3 (NAC-β) genes, so it is possible that different NAC complexes direct the transport of different proteins.

WASP was identified as the protein responsible for a chromosome X-linked recessive immunodeficiency disorder known as the Wiskott–Aldrich Syndrome [34]. It was found to interact with the Nck protein though Src homology III domains [35] and with the Cdc42 protein through a G-protein-binding domain [36,37]. Nck is believed to be an adapter protein in signal transduction as it interacts with a protein kinase, whereas Cdc42 is a member of the Ras superfamily of small GTPases that regulate diverse cellular functions. This implicates the *S. pombe* WASP-like protein in relating cellular signals, perhaps in metal stress. This mutant is also hypersensitive to Ca^{2+}. Both Ca^{2+} and cAMP are known to be secondary messengers in signal transduction and a high Ca^{2+} level reduces the level of cAMP. When cAMP is provided in the medium, the mutant becomes insensitive to Ca^{2+} or to Cd^{2+}. Since cAMP is known to modulate the activity of protein kinases, it is attractive to postulate that cAMP addition overcomes the mutation by affecting a step downstream of the signal transduction pathway. Our current thinking is that the WASP-like protein must be targeted to a designated location, such as to the nucleus, to mediate signal transduction from metal stress. The signal is transduced to elevate the level of cAMP, which then activates a protein kinase to phosphorylate target proteins. In the absence of the associated NAC, the WASP-like protein is missorted to the ER, thus aborting the signal relay (Figure 4).

It is tempting to speculate that one of the target proteins that must be phosphorylated is HMT1. This would provide a plausible means to regulate its activity in the absence of a difference in protein levels, with or without cadmium stress. The regulation of activity through phosphorylation for members of this transport family has precedence. For instance, the cystic fibrosis transmembrane conductance regulator that transports Cl^- ions is regulated by cAMP and phosphorylation. JS237 is also a slow growing mutant, even in the absence of Cd^{2+}. Thus, the mutation probably affects additional targets other than HMT1.

Considerations for plant genetic engineering

The initial discovery and cloning of *hmt1* provided an exciting possibility for engineering high-level expression of this gene. However, the finding that a mannose-1-phosphate guanyltransferase may be involved has raised the possibility that glycosylation may be needed for proper localization of HMT1 or some other component of the Cd-transport pathway. This implies that engineering efforts in reproducing the fission yeast Cd transport system in plants must consider plant glycosylation and sorting of foreign proteins. Given that little is known about protein sorting to the plant vacuolar membrane, this insight has

Figure 4

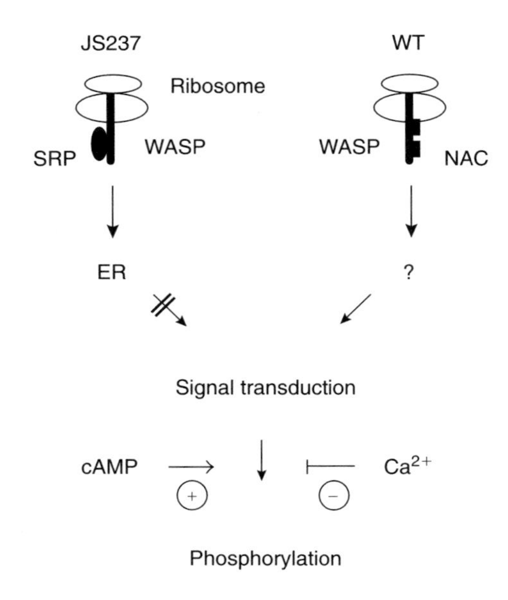

Model of signal transduction in metal response

During Cd stress, the NAC complex (filled rectangles), with this particular BTF3 subunit, binds WASP-nascent polypeptide chains (thick line) formed by the ribosome. The bound complex is protected from the SRP (filled oval). The intracellular location of WASP is probably the nucleus where WASP mediates signal transduction to cause cAMP activation of a protein kinase. In accordance with classical cAMP/Ca^{2+} interactions, a high Ca^{2+} level reduces the level of cAMP. In JS237, which fails to form the high-M_r PCn–CdS complex and is more sensitive to Ca^{2+}, functional BTF3 may not be available to form the NAC complex. Consequently, SRP recognition of WASP directs it to the ER, resulting in the loss of signal transduction and phosphorylation possibly of HMT1 and/or other targets.

redirected efforts to clone the plant homologue of this gene, where signals for proper protein targeting would be present.

The finding that a signal transduction pathway might regulate the metal response, and possibly in the control of the HMT1 transporter, raised another consideration. If phosphorylation is indeed needed for activity, efforts to engineer high-level production of HMT1, its plant homologue or other key proteins in plants may require a way to keep the additional proteins in their transport-active state. If the cell lacks sufficient kinase activity or self-regulates to allow only a certain amount of the HMT1 protein to be in an active conformation, then increased protein production would fail to yield a linear increase in Cd transport. Since this regulatory feature is probably an evolutionary adaptation for energy conservation purposes, it may be possible to isolate mutant proteins that can remain active without phosphorylation.

Should the engineering of high-capacity Cd transport succeed, then the assembly of the stable high-M_r chelate could require commensurately higher sulphide levels. Maximum sequestration could therefore depend on not only enhanced metal transport but also enhanced sulphide production. Elucidating the

Cd-induced sulphide-generating pathway could provide opportunities for genetic manipulation. In contrast to Cd, Pb-induced sulphide production is mediated through the assimilatory sulphate reduction pathway. Hence, the enhancement of PbS formation may be possible through genetic manipulation of this well-described pathway.

A long sought goal in this field has been the cloning of the gene for enzymic synthesis of the peptides. It has been speculated that modification of the cloned gene might lead to hyperproduction of PCns and thereby metal hyperaccumulation. Although the cloning of this gene has yet to be reported, our findings on the FAD/NAD-linked disulphide reductase have revised our thoughts on this prospect. It is possible that hyperproduction of PCns would require a commensurate increase in PCn reduction. Without sufficient reducing power, excess peptides could become oxidized and ineffective for metal binding.

To conclude, research that focuses on the basic aspects of metal detoxification and accumulation provides critical insight in the molecular understanding of the metal response in higher plants. Along with advances in other research areas, these findings will help set the foundation for future engineering efforts in phytoremediation.

References

1. Banuelos, G.S., Cardon, G., Mackey, B., Ben-Asher, J., Wu, L., Beuselinck, P., Akohoue, S. and Zambrzuski, S. (1993) J. Environ. Qual. **22**, 786–797
2. Zayed, A.M. and Terry, N. (1994) J. Plant Physiol. **143**, 8–14
3. Rugh, C.L., Wilde, H.D., Stack, N.M., Thompson, D.M., Summers, A.O. and Meagher, R.B. (1996) Proc. Natl. Acad. Sci. U.S.A. **93**, 3182–3187
4. Moffat, A.S. (1995) Science **269**, 302–303
5. Baker, A.J.M. and Brooks, R.R. (1989) Biorecovery **1**, 81–126
6. Cunningham, S. and Ow, D.W. (1996) Plant Physiol. **110**, 715–719
7. Ow, D.W. (1993) *In Vitro* Cell. Dev. Biol. **29P**, 213–219
8. Hamer, D.H. (1986) Annu. Rev. Biochem. **55**, 913–951
9. Misra, S. and Gedamu, L. (1989) Theor. Appl. Genet. **78**, 161–168
10. Yeargan. R., Maiti, I.B., Nielsen, M.T., Hunt, A.G. and Wagner, G.J. (1992) Transgenic Res. **1**, 261–267
11. Elmayan, T. and Tepfar, M. (1994) Plant J. **6**, 433–440
12. Hattori, J., Labbe, H. and Miki, B.L. (1994) Genome **37**, 508–512
13. Pan, A., Tie, F., Duau, Z., Yang, M., Wang, Z., Li, L., Chen, Z. and Ru, B. (1994) Mol. Gen. Genet. **242**, 666–674
14. Robinson, N.J., Tommey, A.M., Kuske, C. and Jackson, P.J. (1993) Biochem. J. **295**, 1–10
15. Steffens, J.C. (1990) Annu. Rev. Plant Physiol. Plant Mol. Biol. **41**, 553–575
16. Rauser, W. (1995) Plant Physiol. **109**, 1141–1149
17. Grill, E., Winnacker, E.L. and Zenk, M.H. (1987) Proc. Natl. Acad. Sci. U.S.A. **84**, 439–443
18. Thumann, J., Grill, E., Winnacker, E.-L. and Zenk, M.H. (1991) FEBS Lett. **284**, 66–69
19. Mehra, R.K., Kodati, V.R. and Abdullah, R. (1995) Biochem. Biophy. Res. Commun. **215**, 730–736
20. Mehra, R.K., Miclat, J., Kodati, V.R., Abdullah, R., Hunter, T.C. and Muchandani, P. (1996) Biochem. J. **314**, 73–82
21. Maitani, T., Kubota, H., Sato, K. and Yamada, T. (1996) Plant Phsyiol. **110**, 1145–1150
22. Ortiz, D.F., Kreppel, L., Speiser, D.M., Scheel, G., McDonald, G. and Ow, D.W. (1992) EMBO J. **11** , 3491–3499
23. Ortiz, D.F., Ruscitti, T., McCue, K. and Ow, D.W. (1995) J. Biol. Chem. **270**, 4721–4728
24. Salt, D.E. and Rauser, W.E. (1995) Plant Physiol. **107**, 1293–1301
25. Salt, D.E. and Wagner, G.J. (1993) J. Biol. Chem. **268**, 12297–12302
26. Stasdeit, H., Duhme, A.-K., Kneer, R., Zenk, M.H., Hermes, C. and Nolting, H.-F. (1991) J. Chem. Soc., Chem. Commun. **16**, 1129–1130
27. Dameron, C.T., Reese, R.N., Mehra, R.K., Kortan, A.R., Carroll, P.J., Steigerwald, M.L., Brus, L.E. and Winge, D.R. (1989) Nature (London) **338**, 596–597
28. Reese, R.N. and Winge, D.R. (1988) J. Biol. Chem. **263**, 12832–12835

29. Speiser, D.M., Ortiz, D.F., Kreppel, L., Scheel, G., McDonald, G. and Ow, D.W. (1992) Mol. Cell. Biol. **12**, 5301–5310
30. Juang, R.-H., McCue, K.F. and Ow, D.W. (1993) Arch. Biochem. Biophys. **304**, 392–401
31. Porter, D.J.T., Rudie, N.G. and Bright, H.J. (1983) Arch. Biochem. Biophys. **225**, 157–163
32. Zheng, X.M., Black, D., Chambon, P. and Egly, J.M. (1990) Nature (London) **344**, 556–559
33. Wiedmann, B., Sakai, H., Davis, T.A. and Wiedmann, M. (1994) Nature (London) **370**, 434–440
34. Derry, J.M., Ochs, H.D. and Francke, U. (1994) Cell **78**, 635–644
35. Rivero-Lezcano, O.M., Marcilla, A., Sameshima, J.H. and Robbins, K.C. (1995) Mol. Cell. Biol. **15**, 5725–5731
36. Symons, M., Derry, J.M.J., Karlak, B., Jiang, S., Lemahieu, V., McCormick, F., Francke, U. and Abo, A. (1996) Cell **84**, 723–734
37. Kolluri, R., Tolias, K.F., Carpenter, C.L., Rosen, F.S. and Kirchhausen, T. (1996) Proc. Natl. Acad. Sci. U.S.A. **93**, 5615–5618

Production of the biodegradable plastic polyhydroxyalkanoates in plants

Kieran M. Elborough*, Andrew J. White, Steven Z. Hanley and Antoni R. Slabas
Department of Biological Sciences, University of Durham, South Road, Durham DH1 3LE, U.K.

Abstract

Polyhydroxyalkanoates (PHAs) represent a potential source of biodegradable polymers that could occupy a niche presently occupied by plastics made from petrochemicals. Significant levels of polyhydroxybutyrate (PHB) have been produced in the model plant *Arabidopsis*, with levels of up to 14% in the leaf [1,2]. *Arabidopsis*, however, whilst related to oil-seed rape is not a crop of agronomic value in its own right. Using specific protein targeting to seeds and plastid expression we are currently evaluating the potential to produce high levels of seed-specific expression of PHAs, which is required for a commercially viable product.

Introduction

The use of petrochemicals cannot be sustained for modern day living by the earth's oil deposits indefinitely. Modern chemical science has allowed the production of a wide range of synthetic products from this valuable resource that have become essential for modern day living. However, since the products are synthetic they are not usually biodegradable. Our changing view on the use of environmentally friendly products and our realization that oil is not an infinite resource, has directed us to look for alternative methods and sources for producing plastic. Ideally, society would benefit from the availability of disposable plastic products that have the same desirable properties as the petrochemical products, but are made from a renewable resource and are truly biodegradable for disposal. The production of such materials by chemical synthesis from oil-based feedstocks has met with problems in achieving the required molecular mass and stereospecificity during chemical synthesis. In addition it is thought that such synthetic materials would not be seen by the consumer as 'natural' or renewable and would not be able to compete with renewable alternatives derived from bacteria and plants.

*To whom correspondence should be addressed.

Polyhydroxyalkanoates

Polyhydroxyalkanoates (PHAs) are high-molecular-mass aliphatic polyoxoesters produced as carbon-storage compounds by many bacteria and deposited in inclusion bodies. The most characterized PHA biosynthesis pathway to date is from the Gram-negative bacterium *Alcaligenes eutrophus* [3], but there are comparable pathways in many prokaryotes, including purple and green non-sulphur photosynthetic bacteria, purple sulphur bacteria and pseudomonads (see [4]). The polymers are synthesized from simple metabolic intermediates as a store of excess carbon or as a sink for reducing equivalents. Polymerization allows large amounts of carbon to be stored at minimal cost without disrupting the osmotic balance of the cell.

In several organisms, PHAs are used as energy sources during the final stages of sporulation under conditions of minimal carbon in the growth medium. Many organisms from a variety of different environments have been shown to possess the depolymerases that are required for PHA breakdown. It is for this reason that PHAs are entirely biodegradable in a range of environmental conditions. Many aerobic and anaerobic bacteria and fungi are known to break down PHAs by secreting extracellular depolymerases, producing water-soluble monomers and oligomers suitable as carbon sources [5]. These organisms are present in soil, activated sludge, lake water, sea water, estuarine sediment and anaerobic sludge [6,7]. The rate of degradation is dependent upon the organism population and temperature. Complete degradation is always achieved with the concomitant production of carbon dioxide and water in aerobic environments and methane in anaerobic environments.

Small quantities of a short low-molecular-mass polyhydroxybutyrate (PHB) polymer (100–200 units) have been found in many prokaryotes and eukaryotes (including human blood plasma – see [8]) where is seems to function as a membrane ion channel or protein carrier [5,9].

Bacterial synthesis of PHAs

The basic mechanism of synthesis of PHAs involves the polymerization of hydroxyalkanoic acid thioesters (originally elucidated by Merrick and Doudoroff [10]). The feedstock chemical for PHB synthesis in *Alcaligenes eutrophus* is acetyl CoA.

Three genes are responsible for PHB synthesis in *A. eutrophus* [3]. The first enzymic step involves a homotetramer 3-ketothiolase (encoded by the gene *phbA*), which condenses two acetyl-CoA molecules to form acetoacetyl-CoA. The second homotetramer enzyme, acetoacetyl-CoA reductase (encoded for by the *phbB* gene) utilizes NADPH in a reductive step to produce D-(–)-3-hydroxy-butyryl-CoA. The third protein, PHB synthase, is encoded by the *phbC* gene. The first two enzymes are soluble, the last only in the presence of PHB. Gerngross and co-workers [11] used immunogold techniques to show that PHA synthase is exclusively associated with the surface (not internally) of PHA granules under conditions of carbon accumulation. Steinbüchel and Schlegel [3] cloned the *phbA*, *phbB* and *phbC* genes from *A. eutrophus* and demonstrated that they were organized into a single operon in the order CAB. All three enzymes are constitutively expressed and there appear to be no transcriptional or translational controls.

A four-step pathway for the synthesis of the PHB precursor, followed again by a single polymerization step, is found in *Rhodospirillum rubrum* [12]. The operon in this organism appears to be regulated via sigma factors, explaining the ease with which *Escherichia coli* expression has been achieved [13].

Types and properties of PHA polymers

PHB is tolerated extremely well in mammalian tissues and degrades rapidly via surface attack from micro-organisms [7]. This polymer is 100% stereospecific (i.e. all asymmetric carbons are in the same configuration) and therefore highly crystalline and stiff. It is moisture-resistant, optically pure and piezoelectric. Unfortunately, it decomposes at temperatures just past its melting point (177 °C; [6]) and it can become brittle after only short periods of storage. In addition, while the Young's modulus and tensile strength of PHB are comparable with that of polypropylene, the elongation-to-break is very low (5 as opposed to 400% for polypropylene).

It has been demonstrated in many PHA-synthesizing micro-organisms that the monomer used in the resulting polymer is dependent upon the available carbon source; for example, when propionate is added to the growth medium, polyhydroxyvalerate (PHV) is prominent. There are several possible polymers with monomers of different carbon chain length, which are all aliphatic with varying properties (see [5]). The co-polymers have an even greater range of properties. PHA production in *Alcaligenes* can be manipulated *in vivo* by changing the growth medium to produce co-polymers such as PHB-*co*-PHV. Increasing levels of valerate (via propionate feeding to the bacterial culture) improve the impact strength, flexibility, elongation-to-break and melting temperature (without lowering the degradation temperature). This has been demonstrated in *A. eutrophus* and formed the basis of Zeneca's BIOPOL™ plastic.

The polyhydroxyalkanoate PHD [poly(3-hydroxydecanoate)], derived from a 10-carbon chain monomer, is produced by *Pseudomonas* spp. [14] via a poorly characterized pathway. These organisms can produce a wide variety of polymers, depending upon the substrates made available, although certain organisms produce a polymer that is predominantly PHD. This suggests a preference for the medium chain-length monomers.

If the material can be manufactured economically, the uses of PHAs are varied: packaging films, biodegradable micro-capsules for pharmaceuticals and agrochemicals, disposable health products (razors, nappies, tampons), wound dressings, bone replacements and others [15].

BIOPOL

'BIOPOL™' was Zeneca's commercial trademark for the homopolymer PHB and a series of co-polymers (PHB/V) composed predominantly of PHB together with varying proportions of PHV incorporated randomly throughout the polymer chain. BIOPOL has been produced commercially by fermentation of the bacterium *A. eutrophus* using agriculturally derived glucose as the primary carbon source with propionic acid. 'BIOPOL' is, however, the only thermoplastic that is derived from a natural renewable resource and has the properties of durability, moisture resistance and total biodegradability in a wide range of environments.

Although the uses and advantages of the polymer were obvious there was a key factor limiting the potential market size for 'BIOPOL' , the price!

While commodity plastics derived from petrochemicals market at around £0.7 kg^{-1}, the latest price for BIOPOL was £5–12 kg^{-1}. While scale-up could reduce the costs of the fermentation process, production costs could be drastically reduced by producing the polymer directly in a commonly grown crop. This would remove the high energy and capital costs of the fermentation route used at the moment.

Plant polymers

Crop production of biodegradable plastic offers low capital and energy costs. Modern crops are extremely efficient at biomass accumulation thanks to thousands of years of cultivation. The metabolic diversity of plants also encourages us to look to them in this respect, as they are proven factories for all manner of chemicals. Plant genetic engineering can further enhance these capabilities, as shown by the production of human serum albumin in potatoes, novel fatty acids in oil-seeds, the production of 'plantibodies', and more (see other chapters in this volume).

Although very low levels of a natural PHB polymer have been detected in plants and other eukaryotes the plastic is of low molecular mass and the levels are too low to harvest for plastic production. However, with the fast developing science of molecular biology it has been possible to genetically engineer plants to utilize the genetic information needed for the biosynthetic pathway from bacteria that produce PHB.

The cost of producing PHA polymers in oil-seeds (termed PHYTOPOL) is dependent upon two factors – the amount of polymer stored in the plant, and the costs of the extraction process. A cost of £1 kg^{-1} could be within reach, creating the opportunity for 'PHYTOPOL' to be used economically in packaging, agricultural and medical applications. Expression in oil-seeds offers an opportunity for the economic production of other PHA polymers with different characteristics. PHD synthesized by pseudomonads [16,17], is a thermoplastic elastomer and a potential biodegradable substitute for polyurethane, and could potentially be produced in rape. The wide range of alkanoic acid monomers produced by oil-seeds provides further opportunities for incorporating functional chemical groups into the polymers, allowing down-stream processing to produce a whole new range of plastics.

The model experimental plant *Arabidopsis thaliana* was first successfully transformed with the *A. eutrophus* cDNAs encoding the reductase and synthase enzymes [18]. The ketothiolase was known to be already present in plant cells, acting in the isoprenoid biosynthesis pathway, and was therefore not required for the transformation experiment. The plants showed the expected production of PHB in the cell cytoplasm but at a relatively low level of 0.02%. Poirier and co-workers [18,19] used the vector pBI121 to insert the *phbB* and *phbC* genes into the genomes of separate lines of plants. Activity of *phbC*-derived protein could not be tested without the presence of the other gene products. After crossing the transgenic lines, which resulted in some activity losses, PHB inclusion bodies could be seen in the cytoplasm, vacuole and nucleus. The growth and fertility of these hybrids were considerably reduced, possibly due to acetyl-CoA depletion (disrupting the mevalonate pathway and hence affecting the production of terpene-based metabolites such as carotenoids, sterols, hormones, flavonoids, quinones and lipids) or through interference with nuclear processes.

The same group addressed the problem of cytosolic toxicity by expressing the PHA genes in the plastid [1]. The plastid had already been shown to accommodate large amounts of starch without any detrimental effect. In addition, the plastids had already been shown to be the site of *de novo* fatty acid biosynthesis, utilizing the same feedstock, acetyl-CoA, as the PHB biosynthetic pathway. Nawrath and co-workers [1] therefore engineered a RUBISCO small subunit plastid transit peptide (plus some RUBISCO small subunit sequence) upstream of three pBI121-*phb* constructs (including the *phbA* cDNA as the isopentenyl pyrophosphate pathway is not present in differentiated plastids) and transformed *Arabidopsis thaliana*. The use of Western blots confirmed the independent expression of all three proteins in transgenic lines, which were then crossed, which again resulted in the loss of some activity.

The acetyl-CoA flux in the plastids is higher than in the plant cytoplasm and the vigour of the transgenics was equal to the wild-type lines. Yields of PHB were up to 14% of the plant dry weight after 60 days growth [1].

Strategies to increase or optimize PHB levels in plants

We have generated antisense plants that are down-regulated at two key steps in fatty acid synthesis: (1) acetyl-CoA carboxylase and (ii) β-ketoreductase [20,21]. Plant acetyl-CoA carboxylase [EC 6.4.1.3] is one of the pivotal enzymes of fatty acid biosynthesis in both seed and leaf tissues and is thought to be an important regulatory step in the *de novo* synthesis of fatty acids in the chloroplast [22]. By inhibiting these enzyme activities it is thought that the levels of acetyl-CoA may be raised in the plastid. Acetyl-CoA is the feedstock for PHB synthesis. The corresponding cDNAs were cloned from oil-seed rape [23,24] and manipulated to act as antisense constructs in oil-seed rape plants [20]. When homozygous these plants will be analysed and used to cross with PHB-producing plants. The antisense genes are linked to both a 'constitutive' CaMV35S promoter and a seed-specific promoter from the acyl carrier protein (ACP) gene of *Brassica napus*. The ACP promoter, isolated from oil-seed rape [25], shows both tissue and temporal specificity during oil-seed rape embryogenesis.

References

1. Nawrath, C., Poirier, Y. and Somerville, C. (1994) Targeting of the polyhydroxybutyrate biosynthetic pathway to the plastids of *Arabidopsis thaliana* results in high levels of polymer accumulation. Proc. Natl. Acad. Sci. U.S.A. **91**, 12760–12764
2. Nawrath, C., Poirier, Y. and Somerville, C. (1995) Plant polymers for biodegradable plastics: cellulose, starch and polyhydroxyalkanoates. Mol. Breed. **1**, 105–122
3. Steinbüchel, A. and Schlegel, H.G. (1991) Physiology and molecular genetics of poly(β-hydroxyalkanoic acid) synthesis in *Alcaligenes eutrophus*. Mol. Microbiol. **5**, 535–542
4. Steinbüchel, A., Hustede, E., Liebergesell, M., Pieper, U., Timm, A. and Valentin, H. (1992) Molecular basis for biosynthesis and accumulation of polyhydroxyalkanoic acids in bacteria. FEMS Microbiol. Rev. **103**, 217–230
5. Lee, S.Y. (1996) Bacterial polyhydroxyalkanoates. Biotechnol. Bioeng. **49**, 1–14
6. Luzier, W.D. (1995) Materials derived from biomass/biodegradable materials. Proc. Natl. Acad. Sci. U.S.A. **89**, 839–842
7. Doi, Y. (1995) Microbial biosynthesis, physical properties, and biodegradability of polyhydroxyalkanoates. Macromol. Symp. **98**, 585–599
8. Seebach, D., Brunner, A., Bürger, H.M., Schneider, J. and Reusch, R.N. (1994) Isolation and ^{1}H-NMR spectroscopic identification of poly(3-hydroxybutanoate) from prokaryotic and eukaryotic organisms. Eur. J. Biochem. **224**, 317–328
9. Reusch, R.N. (1995) Low molecular weight complexed poly(3-hydroxybutyrate): a dynamic and versatile molecule *in vivo*. Can. J. Microbiol. **41**(Suppl. 1), 50–54
10. Merrick, J.M. and Doudoroff, M. (1961) Enzymatic synthesis of poly(3-hydroxybutyric acid in bacteria. Nature (London) **189**, 890–892
11. Gerngross, T.U., Reilly, P., Stubbe, J., Sinskey, A.J. and Peoples, O.P. (1993) Immunocytochemical analysis of poly(3-hydroxybutyrate) (PHB) synthase in *Alcaligenes eutrophus* H16: localization of the synthase enzyme at the surface of PHB granules. J. Bacteriol. **175**, 5289–5293
12. Moskowitz, G.J. and Merrick, J.M. (1969) Metabolism of of poly-(3-hydroxybutyrate). II. Enzymatic synthesis of D-(–)-β-hydroxybutyryl coenzyme A by an enoyl hydrase from *Rhodospirillum rubrum*. Biochemistry **8**, 2748–2755
13. Lee, S.Y. and Chang, H.N. (1995) Production of poly(3-hydroxybutyric acid) by recombinant *Escherichia coli* strains – genetic and fermentation studies. Can. J. Microbiol. **41**(Suppl. 1), 207–215
14. Timm, A. and Steinbüchel, A. (1992) Cloning and molecular analysis of the poly(3-hydroxyalkanoic acid) gene locus of *Pseudomonas aeruginosa* PAO1. Eur. J. Biochem. **209**, 15–30
15. Brandl, H., Bachofen, R., Mayer, J. and Wintermantel, E. (1995) Degradation and applications of polyhydroxyalkanoates. Can. J. Microbiol. **41**(Suppl. 1), 143–153
16. Eggink, G., De Waard, P. and Huijberts, G.N.M. (1992) The role of fatty acid biosynthesis and degradation in the supply of substrates for poly(3-hydroxyalkanoate) formation in *Pseudomonas putida*. FEMS Microbiol. Rev. **103**, 159–164
17. Eggink, G., De Waard, P. and Huijberts, G.N.M. (1995) Formation of novel poly(hydroxyalkanoates) from long-chain fatty acids. Can. J. Microbiol. **41**(Suppl. 1), 14–21
18. Poirier, Y., Dennis, D.E., Klomparens, K. and Somerville, C. (1992) Polyhydroxybutyrate, a biodegradable thermoplastic, produced in transgenic plants. Science **256**, 520–523
19. Poirier, Y., Dennis, D.E., Klomparens, K., Nawrath, C. and Somerville, C. (1992) Perspectives on the production of polyhydroxyalkanoates in plants. FEMS Microbiol. Rev. **237**, 237–246
20. White, A.J., Markham, J.E., Slabas, A.R. and Elborough, K.M. (1997) in Physiology, Biochemistry and Molecular Biology of Plant Lipids (Williams, J.P., Khan, M.U. and Lem, N.W., eds.), Kluwer Academic, Dordrecht
21. Elborough, K.M., Markham, J.E., White, A.J. and Slabas, A.R. (1997) in Biotin carboxyl carrier protein and biotin carboxylase subunits of the multi subunit form of acetyl CoA carboxylase from *Brassica napus* (Williams, J.P., Khan, M.U. and Lem, N.W., eds.), Kluwer Academic, Dordrecht
22. Post-beittenmiller, D., Roughan, G. and Ohlrogge, J.B. (1992) Regulation of plant fatty-acid biosynthesis – analysis of acyl-CoA and acyl-acyl carrier protein substrate pools in spinach and pea-chloroplasts. Plant Physiol. **100**, 923–930

23. Elborough, K.M., Swinhoe, R., Winz, R., Kroon, J.T.M., Farnsworth, L., Fawcett, T., Martinez-Rivas, J.M. and Slabas, A.R. (1994) Isolation of cDNAs from *Brassica napus* encoding the biotin-binding and transcarboxylase domains of AcetylCoA Carboxylase: assignment of the domain structure in a full length *Arabidopsis thaliana* genomic clone. Biochem. J. **301**, 599–605
24. Elborough, K.M., Winz, R., Deka, R.K., Markham, J.E., White, A.J., Rawsthorne, S. and Slabas, A.R. (1996) Biotin carboxyl carrier protein and carboxyl transferase subunits of the multi subunit form of acetyl CoA carboxylase from *Brassica napus*: cloning and analysis of expression during oilseed rape embryogenesis. Biochem. J. **315**, 103–112
25. deSilva, J., Robinson, S.J. and Safford, R. (1992) The isolation and functional-characterization of a *Brassica napus* acyl carrier protein-5′ flanking region involved in the regulation of seed storage lipid-synthesis. Plant Mol. Biol. **18**, 1163–1172

Manipulating metabolic pathways in cotton fibre: synthesis of polyhydroxybutyrate

Maliyakal E. John
Agracetus, a unit of Monsanto, 8520 University Green, Middleton, WI 53562, U.S.A.

Abstract

Metabolic pathway engineering in cotton has been accomplished by introducing modified *Alcaligenes eutrophus* genes encoding the enzymes acetoacetyl-CoA reductase and polyhydroxyalkanoate (PHA) synthase. These two enzymes, when produced in fibre cells along with endogenous β-ketothiolase, catalyse the polymerization of acetyl-CoA to polyhydroxybutyrate (PHB). Cotton plants not containing the transgenes did not produce PHB. The majority (66%) of PHB in cotton fibre was of $M_r = 0.6 \times 10^6$ or more and was stable throughout fibre development. The chemical identity of PHB in fibres was confirmed by HPLC and GC-mass spectrometry. The rates of heat uptake and cooling were slower in transgenic fibres when compared with the controls. Thus, the transgenic fibres exhibited better insulating properties. This is the first demonstration of fibre trait modification through genetic engineering.

Introduction

Plant-derived natural fibres of industrial importance are cotton, silk, linen and jute. The projected world production of cotton, the pre-eminent natural fibre, was 19.9 million metric tons in 1996 [1]. The United States, China and India account for 55% of the world's cotton production. Because of its aesthetically pleasing appearance and feel, cotton is prized for textile applications. The present estimation of cotton textile use in 1996 is 2.16×10^9 kg [2]. The major competition for cotton comes from man-made fibres that are either derived from naturally occurring cellulose (rayon, acetate and triacetate) or petrochemicals (nylon, polyester and acrylic). Rapid advancements in chemical and manufacturing technologies have produced tremendous improvements in the quality and variety of synthetic fibres in recent years. Examples are superior dyeability, appearance, strength, feel and stain and moisture resistance. To remain competitive in textile markets, the cotton fibre industry needs to lower production costs, increase quality, incorporate new fibre traits and find new product applications. All of these objectives may be attainable by applying recombinant DNA technology to cotton. As an example, the amount of insecticide application can be reduced by expressing an insecticidal protein toxin gene derived from *Bacillus thuringiensis* in cotton [3,4]. Commercial application of this technology was achieved by Monsanto Co. (St. Louis, MO, U.S.A.) in 1996. Similar recombinant DNA technology may be applied to enhance or add new traits to fibres [5].

Cotton fibre or seed hair is a terminally differentiated single epidermal cell made up of two walls, primary and secondary. The primary wall (0.1 to 0.2 μm thick) is made up of cellulose (30%), other neutral polysaccharides, waxes, pectic compounds and proteins [6]. The secondary wall (8 to 10 μm thick) is made up of cellulose. The primary wall is synthesized during the two early stages of fibre development, the initiation and primary wall synthesis stages (0–20 days post-anthesis; DPA), while the secondary wall formation occurs during the third stage of fibre development (16–45-DPA). Maturation, the final stage of fibre development (45–50-DPA), is associated with changes in mineral content and protein levels. Unprocessed mature cotton fibre is 89% cellulose [7]. The chemical composition and microstructure of the primary and secondary walls dictate the chemical reactivity, thermal characteristics, water absorption and various physical properties of the fibre. Many of these properties are important for the manufacturing of textile products. Thus, it is necessary for genetically engineered fibre to retain all of the desirable properties of cotton while enhancing existing properties or adding new traits.

One of the strategies being considered to modify fibre properties is to synthesize a new biopolymer in cotton. Depending upon the biopolymer selected and its level, the new polymer may impart desirable traits to the cotton fibre. As a model system to test this concept, we selected a PHA, poly-D(–)-3-hydroxybutyrate (PHB) for synthesis in fibre. PHAs are biodegradable thermoplastic compounds with similar chemical and physical properties to polypropylene [8,9]. PHAs are produced by many bacteria as inclusion bodies and serve as carbon sources and electron sinks [10]. *Alcaligenes eutrophus* produces bioplastics (PHB) from acetyl-CoA through a three-enzyme pathway. Two molecules of acetyl-CoA are joined by β-ketothiolase to form acetoacetyl-CoA that is then reduced by acetoacetyl-CoA reductase to (*R*)-(–)-3-hydroxybutyryl-CoA. This activated monomer is then polymerized by PHA synthase to form PHB. The production of PHB in transgenic plants has been achieved, with the objective of large-scale isolation [11].

Materials and methods

Cotton plants (*Gossypium hirsutum* L. cv DP50) were grown in a greenhouse. The protocols for RNA, DNA and protein isolation from fibres were described previously [12]. Protocols for northern and Southern analyses, plasmid subcloning and other standard molecular biology techniques have been described by Ausubel and co-workers [13]. PCR was used to amplify the coding region of the *phaB* gene [14,15]. The detailed descriptions of cloning and the construction of expression vectors have been given [15]. The *phaB* gene was linked to a poly(A) addition signal (280 bp) of *Agrobacterium* nopaline synthase gene at the 3′ end and ligated to the cotton promoter E6 [12]. A cauliflower mosaic virus 35s promoter linked to the β-glucuronidase (*GUS*) gene was added to the above plasmid to generate the cotton expression vector, pE6-B. A similar construct with a second cotton promoter, FbL2A, was also generated (pFbL2A-B, [16]). The E6 and FbL2A promoter fragments contained 33 and 44 bp of untranslated 5′ leader sequences, respectively, and translation initiation is expected to occur at the first ATG codon of the *phaB* gene. The *phaC* gene coding region (1770 bp, [17]) was amplified from *A. eutrophus* DNA as two fragments that were ligated together to form the complete gene [15]. The sequence of the insert was determined. The *phaC* insert was also cloned into the expression vector containing the 35s promoter and the Nos poly(A) addition signal to generate p35-C plasmid [15]. Enzymic assays for thiolase were performed by the method of Nishimura and co-workers [18]. Acetoacetyl-CoA reductase activity was measured by a NADPH-dependent spectrophotometric assay as described by Saito and co-workers [19], and that of PHA synthase by a radiometric assay with [^{3}H]D-(–)-3-hydroxy-butyryl-CoA [20].

Particle-bombardment-mediated cotton transformation

The chimaeric genes, pE6-B (or pFbL2A-B) and p35-C, were introduced into cotton by particle bombardment as described in detail by McCabe and Martinell [21]. In brief, surface-sterilized DP50 seeds were germinated and the embryonic axes were removed. The meristem of each seed axis was exposed by dissection under a microscope and incubated in MS medium overnight at 15 °C in the dark. The meristems were subjected to particle bombardment using an *Accell*® electrical discharge device described by Christou and co-workers [22] (*Accell*® is the trademark of the electric device for particle bombardment of Agracetus). Plasmids were precipitated on to 1.5–3.0 μm gold beads (Alfa Chemicals Co.) and loaded on to 18 × 18 mm squares of aluminized Mylar (DuPont 50MMC) carrier sheets at a rate of 0.05 mg of gold + DNA complex per cm^2. The carrier sheets were then accelerated towards the cotton seed axes by the discharge of an 18 kV arc [21]. The seed axes were then allowed to germinate and grow for four to six weeks prior to testing the leaves for *GUS* expression [23]. The resulting plants were chimaeric for the input genes. Each leaf was tested for GUS activity and selectively pruned to isolate nodes or axillary buds subtending the transformed leaves. The process was continued until a transformed plant was obtained. The transformation process resulted in either epidermal or germline transformants. In the former, only the epidermis was transformed and the transgene was not transmitted to its seeds. The germline transformants, however, passed the transgene to their progeny in a Mendelian fashion [21]. Cotton fibres are epidermal cells and therefore both epidermal or germline transformants are useful in evaluating fibre modifications.

Detection of PHB and measurement of thermal properties

Transgenic fibres were treated with Nile Blue A fluorescent stain and examined at excitation wavelengths of 546 nm [24]. PHB granules were also detected by transmission electron microscopy analysis (University of Wisconsin, Madison, WI, U.S.A.) [25]. PHB was extracted and converted into butenoic acid and detected by HPLC [26]. Gas chromatography–mass spectrometry (GC–MS) analysis of the ethyl ester derivative of PHB was done as described by Findlay and White [27] at the Chemistry Department, University of Wisconsin.

Thermal conductivity (TC), thermogravimetric analysis (TGA), specific heat, and differential scanning calorimetry (DSC) measurements of cotton fibres were undertaken by MATECH Associates (Scranton, PA, U.S.A.) using TA instruments (TA Instruments Inc., New Castle, DE, U.S.A.). Thermal analysis protocols conformed with those recommended by the International Confederation for Thermal Analysis [28].

Results and discussion

The biopolymer selected for synthesis in cotton fibre is PHB, one of the many PHAs produced by bacteria. Its biosynthetic pathway is well understood. Several of the chemical and physical properties of PHAs are sufficiently distinct from those of cellulose and are described in Table 1. Synthesis of PHB in fibre is an example of introducing new enzymes to utilize an existing substrate, acetyl-CoA, to produce a new polymer in fibre.

Metabolic pathway engineering for the synthesis of PHB in cotton requires only the integration of functional *phaB* and *phaC* genes, since β-ketothiolase, which is involved in the synthesis of mevalonate, is ubiquitous in plants. To avoid any detrimental effects of PHB production during plant growth, one of the genes, *phaB*, was engineered for expression in fibre. Reduction in growth and seed production were observed in transgenic *Arabidopsis* as a result of *phaB* overexpression in leaves and other tissues [11]. We introduced *phaB* and *phaC* together with the marker gene *GUS* into cotton by particle bombardment and a total of 85 R0 cotton plants was scored as transgenic, based on *GUS* expression in

Table 1

Properties	Cellulose	PHB	Polypropylene
Biodegradability	Yes	Yes	No
Crystallinity (%)	70–80	65–80	65–70
Melting point (°C)	Chars >250	171–182	176
Tensile strength (MPa)*	500–700	40	39
Solvent resistance	Good	Poor	Good
Moisture regain (65%rh)*	7–8	0.0	0.0
Extension at break (65%rh)*	5–13	6	400
Elastic modulus N/tex (65%rh)*	3.5–7.9	Nk	2.6–4.0
Dye binding	Direct; reactive	Disperse	Acidic
Thermoplastic†	No	Yes	Yes
Glass transition (°C)		15	−10
Flexibility	Flexible	Brittle	Flexible
Fat resistance‡	Resistant	Permeable	Resistant

**Arthur [7].*

†Poirier and co-workers [30].

‡Hanggi [31].

Comparison of the properties of cellulose, PHB and polypropylene

rh, relative humidity; Nk, not known.

leaves. Fifty-nine of these were epidermal transformants, while the remaining were germline transformants. Table 2 summarizes the results of seed axis bombardment and transformation efficiency. The transformation efficiency was low (0.047%) and therefore required bombardment of a large number of seed axes. The epidermal R0 plants and the R1 progeny of germline transformants were grown and examined for transgene expression. Table 3 shows the range of acetoacetyl-CoA reductase and PHA synthase activities in typical transgenic cotton. Reductase activities varied between a low of 0.05 to a high of 0.5 $\mu mol \cdot min^{-1} \cdot mg^{-1}$ in fibre. PHA synthase activities also show a great variation between transformants. All transgenic cotton that expressed *phaB* and *phaC* showed normal growth and morphology. Figure 1 shows a Western blot of fibre proteins reacted with antibodies for acetoacetyl-CoA reductase and PHA synthase. The reductase antibody reacted with a protein of M_r 26500 in transgenic fibre, whereas control fibre proteins did not. The PHA synthase antibody cross-reacted with a protein of M_r 63000 in transgenic fibres (arrowed in Figure 1B).

Table 2

Exp #	Seed axes (%)	% Survival	Epidermal	Germline	Total
1	6149	59	2 (0.033)	0 (0.0)	2 (0.033)
2	10410	53	5 (0.048)	0 (0.0)	5 (0.048)
3	10238	50	4 (0.04)	3 (0.03)	7 (0.07)
4	7637	35	1 (0.013)	0 (0.0)	1 (0.013)
5	10404	62	1 (0.01)	0 (0.0)	1 (0.01)
6	14159	40	5 (0.04)	4 (0.03)	9 (0.07)
7	15139	35	2 (0.01)	2 (0.01)	4 (0.026)
8	10344	44	1 (0.01)	1 (0.01)	2 (0.02)
9	10130	46	1 (0.01)	1 (0.01)	2 (0.02)
10	10114	44	2 (0.02)	1 (0.01)	3 (0.03)
11	7940	46	4 (0.05)	0 (0.01)	4 (0.05)
12	10262	54	9 (0.09)	1 (0.01)	10 (0.08)
13	5069	46	1 (0.02)	1 (0.025)	2 (0.04)
14	4947	51	4 (0.08)	1 (0.02)	5 (0.1)
15	10034	52	2 (0.02)	0 (0.01)	2 (0.02)
16	7536	53	1 (0.01)	0 (0.0)	1 (0.01)
17	7506	50	3 (0.04)	9 (0.12)	12 (0.16)
18	7513	52	6 (0.08)	1 (0.01)	7 (0.09)
19	7505	53	2 (0.03)	0 (0.0)	2 (0.03)
20	7515	46	3 (0.04)	1 (0.01)	4 (0.05)
Total	180551	49	59 (0.033)	26 (0.014)	85 (0.047)

Particle-bombardment-mediated cotton transformation

The Table summarizes experiments undertaken over a 1 year period . Untreated DP50 seeds were used and transformation was monitored by following GUS expression in leaves and other tissues. Germline events were identified by GUS assays in pollen. Transformation frequencies based on the number of seed axes bombarded are given in parentheses.

Table 3

Plant #	Promoter/genes	Reductase activity (μmol·min^{-1}·mg^{-1})	PHA synthase activity (cpm $\times$ 10^5)	PHB (μg·g of fibre^{-1})
7148*	E6-B/35-C	0.50	1.7	3440
8267*	E6-B/35-C	0.05	0.3	30
6898*	E6-B/35-C	0.16	0.8	1135
6888-7	E6-B/35-C	0.07	ND	30
7271-1	E6-B/35-C	0.07	0.0	–
7271-12	E6-B/35-C	0.07	0.0	–
11707*	FbB/35-C	0.25	1.3	50
10481*	FbB/35-C	0.18	2.8	549
6903*	E6-B/35-C	0.17	0.3	149
5297-5	E6-B/35-C	0.15	0.0	–
5297-31	E6-B/35-C	0.27	0.0	–
DP50 Control		0.0	0.01	0.0

* *Epidermal transformants.*

Expression of acetoacetyl-CoA reductase, PHA synthase and PHB in transgenic cotton fibres

A selected number of transgenic plants expressing bioplastic enzymes and PHB are shown. E6 and Fb refer to the cotton promoters E6 and FbL2A, while 35 is the cauliflower mosaic viral 35s promoter. B and C are acetoacetyl-CoA reductase and PHA synthase genes, respectively. Enzyme levels and PHB contents of fibres are shown. PHB was estimated by HPLC analysis of butenoic acid from the chloroform extracts of 15-DPA fibres (E6 promoter) and 30-DPA fibres (FbL2A promoter).

Figure 1

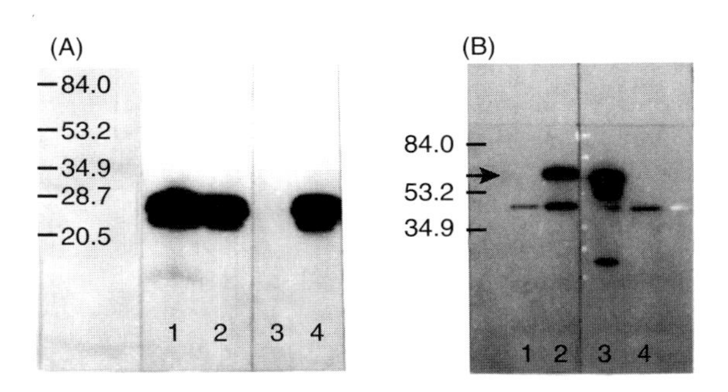

Expression of acetoacetyl-CoA reductase and PHA synthase in transgenic cotton

(A) Western blot analysis of acetoacetyl-CoA reductase in developing cotton fibres. Fibres were isolated at 15-DPA and total proteins extracted. Ten micrograms of total protein were separated by SDS/PAGE (4–20% gradient gel; Novex). The proteins were transferred to a PVDF-Plus membrane and probed with acetoacetyl-CoA reductase antiserum (1:2000 dil.). Cross-reacting proteins were detected with the ECL Western blotting system (Amersham). Exposure time was 5 s. Lanes 1 and 2, transgenic; lane 3, DP50 control; lane 4, A. eutrophus. (B) Western blot analysis of PHA synthase. Experimental conditions were similar to part A except that PHA synthase antiserum was used. The arrow in part B indicates the PHA synthase band cross-reacting with the antibody. Lanes 1 and 4, DP50 control; lane 2, transgenic; lane 3, A. eutrophus.

Developmental regulation of bioplastic genes in fibre

Acetoacetyl-CoA reductase expression in developing fibres was measured in transgenic cotton containing three promoters, a cauliflower mosaic virus promoter (35s), a cotton early fibre promoter (E6) and a cotton late fibre promoter (FbL2A). A comparison of reductase levels in developing fibres of three plants, each carrying different promoter-*phbB* genes, is shown in Figure 2. Both 35s and E6 promoters are active during early fibre development while the FbL2A promoter directs reductase gene activity in late fibre development. The FbL2A promoter also appeared to be stronger than 35s and E6 promoters in fibre [16]. Within each group, the transgenic plants showed a wide variation in enzyme activities. The distribution of 35 transformants based on acetoacetyl-CoA reductase level is shown in Figure 3. Approximately 53% of plants showed moderate enzyme activity (0.1–0.25 μmol·min^{-1}·mg^{-1}) whereas only 5% showed high enzyme activities (>0.5 μmol·min^{-1}·mg^{-1}). The remaining plants showed low (<0.06 μmol·min^{-1}·mg^{-1}) enzyme activities.

The temporal regulation of *phaB* in fibre conforms to the known characteristics of the corresponding E6 and FbL2A promoters [12,15,16]. The optimal levels of reductase and PHA synthase activities, as well as their expression in fibre, are likely to be important for maximum synthesis of PHB. From the limited number of plants studied, it appears that high levels of reductase and PHA synthase levels during early fibre development are conducive for PHB synthesis. Transgenic plants that contained the FbL2A-B gene (active in later fibre development) showed moderately high levels of acetoacetyl-CoA reductase and

Figure 2

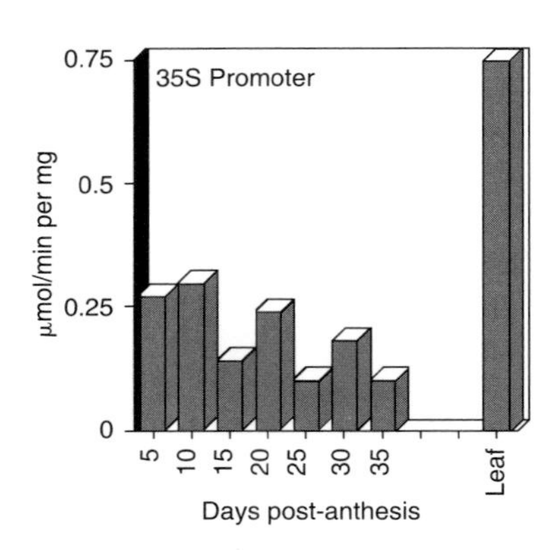

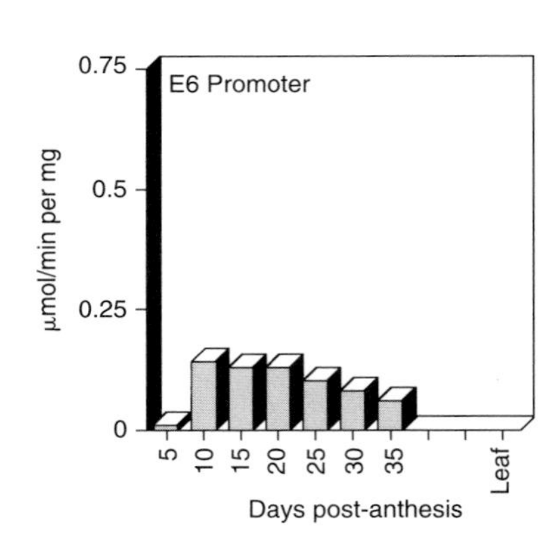

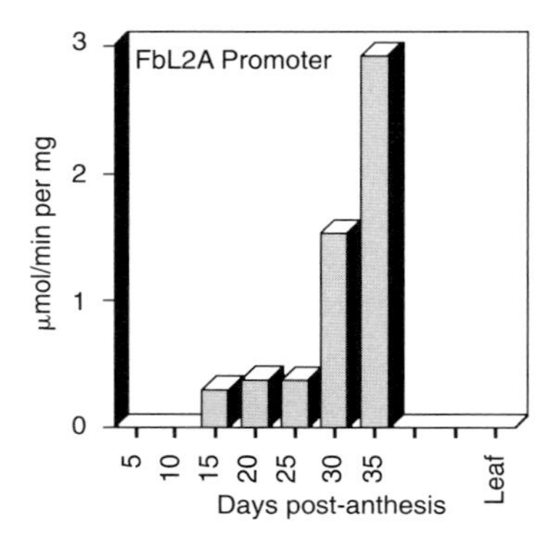

Acetoacetyl-CoA reductase expression in developing fibre cells controlled by 35s, E6 and FbL2A promoters

The patterns shown for each promoter are based on enzyme measurements on a single plant, but were each confirmed in four or more other independent transformants. Reductase expression was also monitored in other tissues. As shown in the Figure, only the 35s-phbB plants showed expression in leaves.

PHA synthase activities, but did not result in high levels of PHB (Table 3). It is possible that decreased levels of acetyl-CoA during late fibre development may be a contributing factor for reduced PHB synthesis [16].

Detection of PHB in transgenic fibres

Epifluorescence microscopy of cells stained with Nile Blue has been used to detect PHB in bacteria as well as in transgenic plants [11,24]. Control and transgenic fibres were stained with Nile Blue A and observed at the excitation wavelength of 546 nm. These experiments revealed fluorescent granules in immature as well as mature transgenic fibres [15]. The control fibres were devoid of fluorescent granules; however, diffused fluorescence was occasionally observed. The transgenic fibres were further subjected to transmission electron microscopy. Cross-sections of fixed immature fibres (30-DPA) of transgenic plants showed

Figure 3

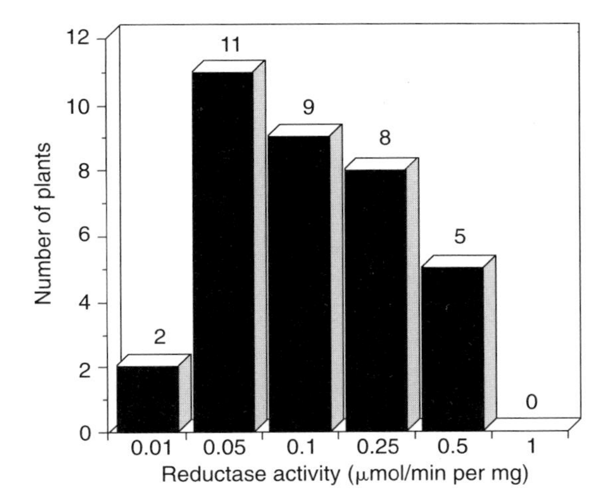

Distribution of acetoacetyl-CoA reductase levels in transgenic cotton

Thirty-five transgenic cotton plants containing the E6 promoter linked to acetoacetyl-CoA reductase were screened for reductase expression in 15-DPA fibres. Plants expressing reductase at 0.006 μmol·min^{-1}·mg^{-1} or higher levels were considered as positive for gene expression and were grouped into classes based on enzyme activity. The first group contained two plants with reductase activity levels in the range 0.006 to 0.01 μmol·min^{-1}·mg^{-1} and the second group (11) in the range of 0.011 to 0.05 μmol·min^{-1}·mg^{-1}. The fifth group had enzyme levels in the range 0.26 to 0.50 μmol·min^{-1}·mg^{-1}.

electron-lucent granules that were in the range of 0.15 to 0.3 μm in diameter. Similar granules were not observed in the control fibre cells (Figure 4).

Further identification of the chemical nature of these granules comes from HPLC and GC–MS studies. Transgenic and control fibres were extracted with chloroform and the extracts were subjected to acid hydrolysis to convert PHB into butenoic acid. Butenoic acid was then detected by HPLC. The retention time for butenoic acid was obtained by subjecting standard butenoic acid (Sigma) to HPLC analysis. Extracts of transgenic fibres, control fibres and bacterial PHB were then analysed by HPLC. The analyses indicated that transgenic fibre extract contained butenoic acid with a similar retention time to that of standard butenoic acid, whereas the control fibre extracts did not (not shown: see [15]). This result suggests that the transgenic fibres contained PHB.

PHB was treated with ethanol/chloroform/hydrochloric acid mixtures for conversion into ethyl ester derivatives that were then analysed by GC–MS [15]. Duplicate experiments were carried out with bacterial PHB. Figure 5 shows the GC profile for bacterial PHB, the control fibre extract and the transgenic fibre extract. The transgenic fibre extract contained a compound with an identical retention time to that of bacterial PHB whereas the control fibre extract did not. Further analysis of the ethyl ester derivatives by GC–MS confirmed that the compound present in the transgenic fibre extract had a similar mass fragmentation pattern to that of the bacterial PHB (Figure 6). Moreover, these two patterns

Figure 4

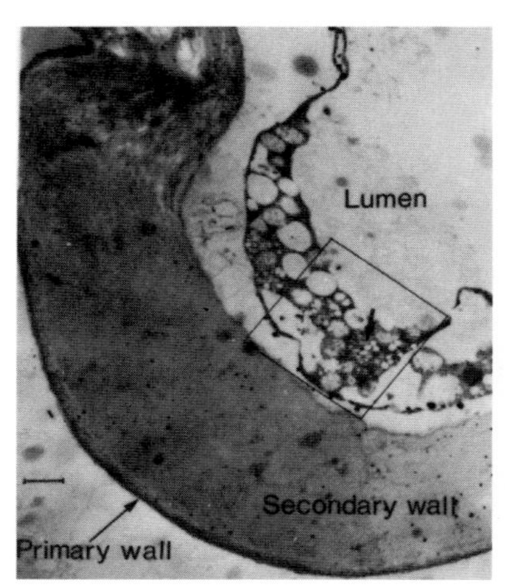

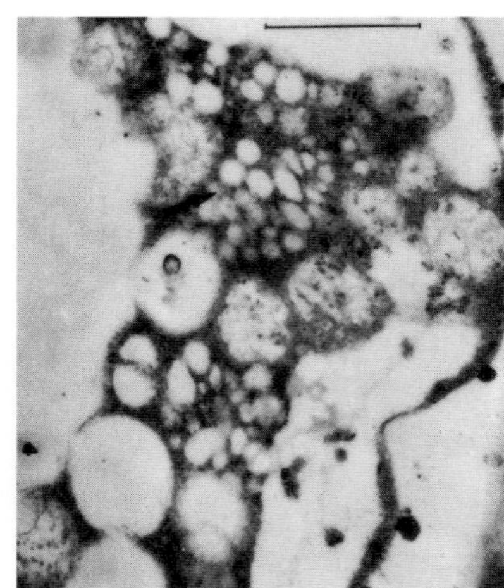

Electron-lucent PHB granules in transgenic fibres

Cotton fibres (#7148, 30-DPA) were fixed with 1% paraformaldehyde, 2% glutaraldehyde in 0.05 M phosphate buffer (pH 7.2) for 2 h at room temperature. After washing and fixing, they were embedded in Spurrs epoxy resin. Seventy-five to 90 nm thick sections were then placed on nickel (300 mesh) grids and stained with 2% uranyl acetate followed by Reynolds lead citrate [15]. Transmission electron (Jeol 100C × II) micrographs of cross-sections of cotton fibres are shown. Arrows indicate clusters of granules in the range of 0.15 to 0.3 mm. Bars represent 1 mm. Reproduced from [15] with permission. Copyright (1996) National Academy of Sciences, U.S.A.

matched with that of the reference compound, ethyl ester hydroxybutyrate. Thus, GC–MS data provide evidence that the transgenic plants produced PHB in fibres [15].

Molecular mass and temporal profile of PHB synthesis in developing fibres

The molecular mass of the PHB polymer in fibres was estimated by gel-permeation chromatography [15]. PHB was fractionated and each fraction converted into butenoic acid and analysed by HPLC. Based on the elution time of M_r standards, a major portion (68.3%), of the PHB in cotton fibre had an M_r of 0.6×10^6 or more, of which 31% had a M_r of 1.8×10^6 or more (not shown: see [15]).

Quantitative estimation of PHB accumulation in developing fibres was undertaken by HPLC analysis. The amounts of PHB from the bolls of different ages from the same plant (#7148) were estimated by HPLC analysis of butenoic acid [Figures 7(A) and 7(B)]. When the amount of PHB was displayed as a function of fibre weight, the PHB level increased up to 15-DPA and then decreased (Figure 7A). It is likely that this decrease is due to the increase in the fibre weight occurring after 15-DPA. Deposition of large quantities of cellulose occurs during the secondary wall synthesis stage of fibre development [29]. When the total weight of PHB during development was estimated per boll, as shown in

Figure 5

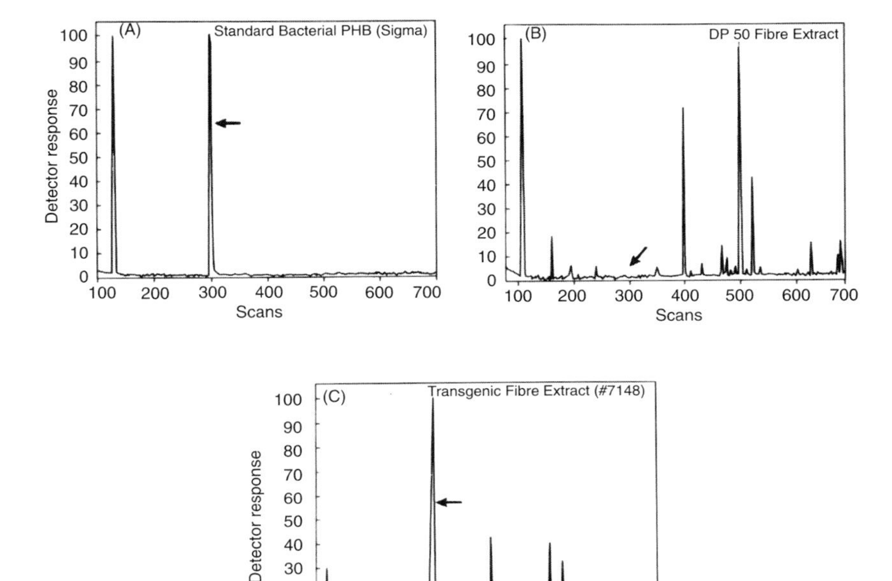

Gas chromatographic separation of β-hydroxy acids

PHB was extracted with chloroform and dried in a N_2 stream. It was then redissolved by heating in 0.5 ml of chloroform for 10 min. at 100 °C. Subsequently, 1.7 ml of ethanol and 0.2 ml of HCl were added and the mixture heated at 100 °C for 4 h. GC analysis was then performed on a Carlo-Erba GC with an electron impact detector and a DB 5 column (50 m; Supelco). Scans were taken at 1 s intervals (700 total). Calibrated range was 17–600 mass units [15]. (A) The GC profile of PHB from A. eutrophus (Sigma). The arrow indicates the PHB peak. (B) GC analyses of the fibre extract from the control DP50 line. The arrow indicates the retention time of PHB from the transgenic #7148 line. (C) GC analysis of transgenic fibre extract. The arrow indicates the ethyl ester of β-hydroxy acid. Reproduced from [15], with permission. Copyright (1996) National Academy of Sciences, U.S.A.

Figure 7(B), no significant decrease was seen. Extraction of PHB from mature fibres is difficult and this fact may account for the small apparent decrease in PHB during fibre maturation (Figure 7B).

Thus, the two bioplastic genes, acetoacetyl-CoA reductase and PHA synthase, when introduced into the cotton genome, result in the production of a new biopolymer in the fibre lumen. The temporal pattern of PHB production is determined by the promoters that drive the bioplastic genes. The chemical and physical properties of PHB in cotton fibres are very similar to those of bacterial PHB.

Effect of PHB on fibre properties

A wide variety of physical and chemical properties of transgenic fibres containing PHB are being investigated. Some of the preliminary results of thermal measurements are described later.

Figure 6

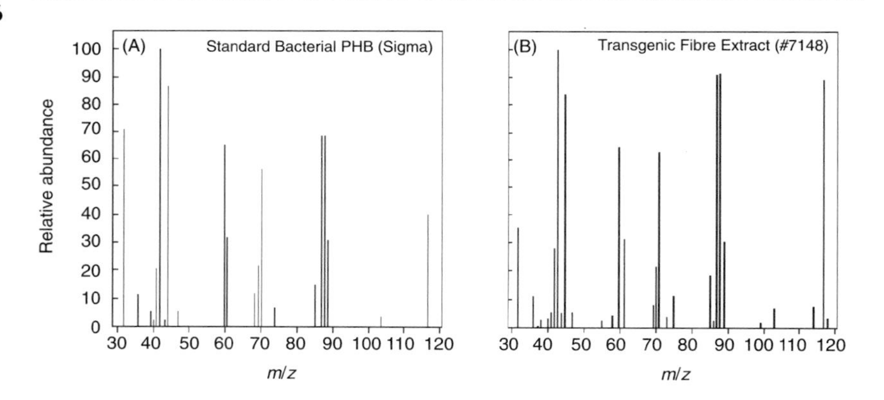

Mass spectroscopic analysis of PHB synthesized in transgenic fibres

The ethyl esters of β-hydroxy acids of bacterial (A) and fibre (B) origins resolved by GC were subjected to impact electron mass spectroscopy performed on a Kratos MS-25 spectrometer. The pattern of mass fragments of the PHB from the transgenic fibre is similar to that of bacterial PHB. The mass fragmentation pattern of reference ethyl ester hydroxybutyrate was similar to those of bacterial and fibre-derived materials (not shown). Adapted from [15].

Figure 7

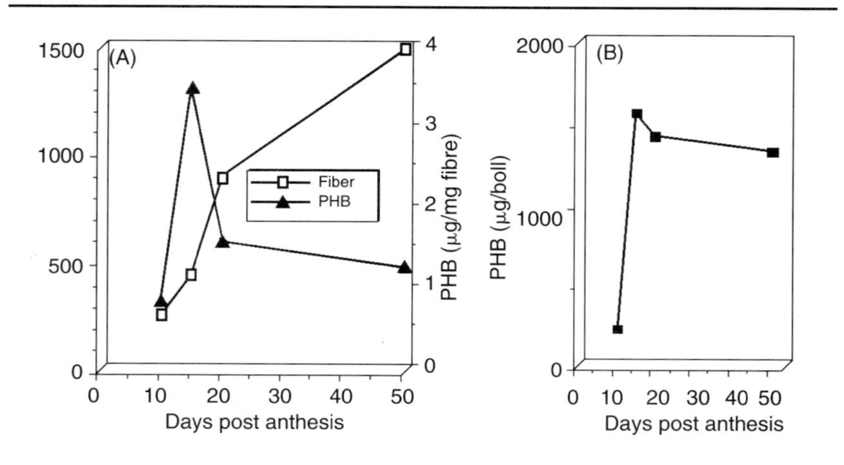

PHB accumulation during fibre development

A, PHB accumulation in relation to fibre weight during development. PHB levels were determined from 10-, 15-, 20- and 50-DPA fibres of #7148. The increase in fibre weight per boll is also shown. The weight of PHB as a function of fibre weight decreased during fibre maturation due to increased fibre weight and not due to degradation of PHB [15]. B, PHB level (total weight per boll) in developing bolls. The level does not decrease significantly during fibre maturation. Reproduced from [15] with permission. Copyright (1996) National Academy of Sciences, U.S.A.

The thermal properties of transgenic fibres were compared with those of control fibres by DSC and TGA analyses. DSC determines qualitative and quantitative heat and temperature transitions by measuring the heat flow rate through the sample. The DSC measurements, when compared with DP50, indicated that the onset of decomposition of transgenic #7148 fibre was advanced. This result was confirmed by TGA analysis. The total heat uptake for #7148 (690.7 $J \cdot g^{-1}$) was 11.6% higher than for DP50 (619.0 $J \cdot g^{-1}$), as shown in Table 4. Two other transgenic samples, #6888-7 containing 30 μg of PHB/g of fibre and #8801 containing 423 μg of PHB/g of fibre, were subjected to heat uptake measurements by DSC. Heat uptake values of 627.5 and 642.3·$J \cdot g^{-1}$ were obtained for #68988-7 and #8801, respectively (not shown). Thus, the heat uptake capacity appears to be related to the amount of PHB present in fibre. The precise relationship between thermal properties and the amounts of PHB in fibre can be established as fibres with greater PHB content are developed. The transgenic fibres (#7148) were spun into yarn by miniature spinning (Starlab VY-5 direct sliver-to-yarn spinning frame) and knitted into cloth by a knitting machine. Unbleached and undyed fabric was then subjected to thermal property measurements by DSC together with the control fabric (DP50). The DSC measurements showed a heat uptake of 695.4 $J \cdot g^{-1}$ for #7148 and 617.8 $J \cdot g^{-1}$ for DP50 fabric, respectively (not shown).

The relative heat transmission capacities of #7148 and DP50 fibres were determined by TC measurements (Table 4). The TC of #7148 fibres (0.264 $W \cdot m^{-1} \cdot °C^{-1}$) was 6.7% lower than DP50 (0.283 $W \cdot m^{-1} \cdot °C^{-1}$), indicating

Table 4

	TGA (% mass loss)	TC (BTU/h/ft 2/c.in)	DSC (heat uptake/ $J \cdot g^{-1}$)	SH (J/g°C)
DP50 control	Total loss = 82.6 Residue = 17.33	1.968	619.0	1.86 (38 °C) 2.692 (60 °C)
#7148	Total loss = 85.55 Residue = 14.45	1.836 (−7%)	690.7 (12%)	2.022 (36 °C) 3.889 (60 °C) (45%)

Thermal properties of transgenic and control fibres

TGA anaylsis indicated that the onset of decomposition for transgenic and control fibres was the same. Weight losses at all stages are higher and faster for transgenic fibres, except for the last stage. The heat flow rate through samples of #7148 and control DP50 fibres was monitored against temperature using a steady-state equilibrium heat flow method. A modified differential scanning calorimeter (DSC: TA Instruments, New Castle, DE, U.S.A.) was used for the thermal conductivity measurements. All conditions for DP50 were identical to those shown for #7148, except: dimension = 6.61 × 2.36 mm and T_2 = 302.18 K. The heat gain/loss characteristics of fibres were measured by DSC. Sample sizes of 10 mg (#7148) and 10.06 mg (DP50) were used. Specific heat (SH) values were obtained for 36 and 60 °C each. A sample size of 10 mg was used in the DP50 control while 10.5 mg was used for #7148. Cell constant values at temperatures of interest were determined by a sapphire standard and zero baseline was established by scanning empty reference and sample pans over the entire temperature range. Cell constant values are E-36 °C = (0.7978 × 10 × 60.83)/(60 × 0.8 × 9.6) = 1.053 and E-60 °C = (0.8432 × 10 × 60.83)/(60 × 0.8 × 10.05) = 1.063. TGA, 10 °C min^{-1} to 700 °C; DSC, 10 °C min^{-1} to 320 °C; TC by steady-state equivalent heat flow (axial/radial); SH, zero baseline of sample sapphire (E-30 °C = 11.053; E-60 °C = 1.063). BTU, British thermal units.

slower cooling down of the material. Thus the DP50 fibres have faster heat dissipation properties.

The heat retention of samples was determined by specific heat measurements at two temperatures (36 and 60 °C). Sample #7148 showed 8.6% higher heat retention than the DP50 sample at 36 °C, while the difference was 44.5% higher for transgenic fibres at 60 °C (Table 4). Thus these results agree with the TC measurements and confirm that sample #7148 has higher heat capacity.

Conclusions

Production of a new polymer, PHB, enhanced the properties of cotton fibre as shown by the studies of thermal characteristics. The transgenic cotton fibres exhibited measurable changes in thermal properties that suggest enhanced insulation characteristics. The transgenic fibres conducted less heat, cooled down slower and took up more heat than conventional cotton fibres. Modified fibres with superior insulating properties may have applications in winter wear or other textile uses where enhanced insulating properties are advantageous. However, the changes in thermal properties are relatively small, as expected from the small amounts of PHB in fibres (0.34% fibre weight in #7148). It is likely that a severalfold increase in PHB synthesis would be required for product applications. Nevertheless, the positive changes in fibre qualities demonstrated here are an indication of the potential of this technology. This study is the first demonstration that synthesis of a new polymer in cotton fibre may enhance fibre traits. As new generations of fibres are developed through genetic engineering they will have impacts on the growth of textile industries that contribute significantly to the economies of many countries.

I acknowledge the expert technical help of Greg Keller, Jennifer Rinehart, Lori Spatola and Michael Petersen. I am thankful to Dr. B. Chowdhury, MATECH Associates, for conducting the thermal studies and for useful discussions. I am grateful to Cheryl Scadlock and Andrea Kersten for editorial assistance. Agreement to the use of A. eutrophus genes was obtained from Metabolix, Inc., Cambridge, MA. Antibodies to PHA synthase and reductase were obtained from S. Pagette of Monsanto (MO). This study was partially funded by the Department of Commerce, National Institute of Technology, Advanced Technology Program grant # 70NANB5H1061.

References

1. Townsend, T.P. (1996) Cotton: Review of the world cotton situation. Monograph by the International Cotton Advisory Committee, Washington, D.C.
2. Textile Fibres: Industry study #420 (1992) The Freedonia Group, Inc., Cleveland, OH
3. Jenkins, J.N. and McCarty, J.C. (1995) in Proceedings Beltwide Cotton Conference, Vol. 1, pp. 171–173, National Cotton Council, Memphis, TN
4. Conner, C. and Deshaies, M. (1996) Cotton Farm. **40**, 8–12
5. John, M.E. (1994) Chem. Ind., 676–679
6. Ryser, U. (1985) Eur. J. Cell. Biol. **39**, 236–256
7. Arthur, J.C. (1990) in Polymers: Fibers and Textiles, a compendium (Kroschwitz, J.I., ed.), pp. 118–141, Wiley, New York
8. Steinbüchel, A. (1991) Acta Biotechnol. **5**, 419–427
9. Muller, H.-M. and Seebach, D. (1993) Angew. Chem., Int. Ed. Engl. **32**, 477–502
10. Steinbüchel, A., Hustede, E., Liebergesell, M., Pieper, U., Timm, A. and Valentin, H. (1992) FEMS Microbiol. Rev. **9**, 217–230

11. Poirier, Y., Dennis, D., Klomparens, K. and Somerville, C. (1992) Science **256**, 520–523
12. John, M.E. and Crow, L.J. (1992) Proc. Natl. Acad. Sci. U.S.A. **89**, 5769–5773
13. Ausubel, F.M., Brent, R., Kingston, R.E., Moore, D.D., Seidman, J.G., Smith, J.A. and Struhl, K. (eds.) (1987) Current Protocols in Molecular Biology, Wiley, New York
14. Peoples, O.P., and Sinskey, A.J. (1989) J. Biol. Chem. **264**, 15293–15297
15. John, M.E. and Keller, G. (1996) Proc. Natl. Acad. Sci. U.S.A. **93**, 12768–12773
16. Rinehart, J.A., Petersen, M.W. and John, M.E. (1996) Plant Physiol. **112**, 1331–1341
17. Peoples, O.P. and Sinskey, A.J. (1989) J. Biol. Chem. **264**, 15298–15303
18. Nishimura, T., Saito, T. and Tomita, K. (1978) Arch. Microbiol. **116**, 21–27
19. Saito, T., Fukui, T., Ikeda, F., Tanaka, Y. and Tomita, K. (1977) Arch. Microbiol. **114**, 211–217
20. Schubert, P., Steinbüchel, A. and Schlegel, H.G. (1988) J. Bacteriol. **170**, 5837–5847
21. McCabe, D.E. and Martinell, B.J. (1993) Biotechnology **11**, 596–598
22. Christou, P., McCabe, D.E., Martinell, B.J. and Swain, W.F. (1990) Trends Biotechnol. **8**, 145–151
23. Jefferson, R.A. (1987) Plant Mol. Biol. Rep. **5**, 387–405
24. Ostle, A.G. and Holt, J.G. (1982) Appl. Environ. Microbiol. **44**, 238–241
25. Hustede, E., Steinbüchel, A. and Schlegel, H.G. (1992) FEMS Microbiol. Lett. **93**, 285–290
26. Karr, D.B., Waters, J.K. and Emerich, D.W. (1983) Appl. Environ. Microbiol. **46**, 1339–1344
27. Findlay, R.H. and White, D.C. (1983) Appl. Environ. Microbiol. **45**, 71–78
28. Hill, J.O. (1991) For Better Thermal Analysis and Calorimetry, 3rd edn., pp. 6–50, International Confederation for Thermal Analysis, New Castle, Australia
29. Meinert, M.C. and Delmer, D.P. (1977) Plant Physiol. **59**, 1088–1097
30. Poirier, Y., Nawrath, C. and Somerview, C. (1995) Biotechnology **13**, 142–150
31. Hanggi, U.J. (1995) FEMS Microbiol. Rev. **16**, 213–220

Lignin manipulation for fibre improvement

Claire Halpin*†, M.E. Knight, A. O'Connell and W. Schuch
Zeneca Plant Sciences, Jealott's Hill Research Station, Bracknell, Berks RG12 6EY, U.K.

Plant fibres provide a cheap and renewable raw material for the production of a variety of valuable products, including textiles and paper. For many such applications, particularly for the production of high quality papers, only the cellulose component of the fibre cell wall is required, and other wall components, such as lignin, have to be mechanically or chemically removed. Lignin is a heterogeneous aromatic polymer with a complex chemical structure that incorporates many condensed bonds and cross-links to other cell wall components [1]. Consequently, lignin removal requires very harsh treatments that are costly both in terms of chemical consumption and environmental impact. It would be of great benefit, therefore, to alter lignin synthesis in plants so that either less is produced or the composition is altered in ways that make it easier to extract. The huge natural variability in the amount and composition of lignins produced by plants [2] suggests that it should be possible to achieve both of these aims without adversely affecting plant health and fitness. Indeed, mutants with altered lignin biosynthesis exist in a number of species [3]. These mutants, characterized by an unusual red–brown coloured lignin and hence known as brown midrib (*bmr*) mutants, show changes in both the amount and composition of lignin produced although the plants are otherwise healthy.

The lignin biosynthetic pathway

Lignin is a product of the phenylpropanoid pathway, which also has branches leading to numerous other important metabolites, including flavonoids, coumarins and phytoalexins [4]. The enzyme phenylalanine ammonia-lyase (PAL) controls entry to the phenylpropanoid pathway and is highly regulated during plant development. It deaminates phenylalanine to cinnamate, which is subsequently hydroxylated to *p*-coumarate by the action of cinnamate 4-hydroxylase (C4H). After PAL and C4H, the pathway is thought to progress through a series of hydroxylation and methylation reactions catalysed by the enzymes cinnamate 3-hydroxylase (C3H), ferulate 5-hydroxylase (F5H) and bi-specific caffeate/5-hydroxyferulate *O*-methyltransferase (COMT), which produce ferulate and sinapate from *p*-coumarate. These three acids are activated by 4-coumarate:CoA ligase (4CL) to the corresponding CoA thioesters. The thioesters undergo two successive reduction reactions catalysed by cinnamoyl-

*To whom correspondence should be addressed.
†Present address: Department of Biological Sciences, University of Dundee, Dundee DD1 4HN, U.K.

CoA reductase (CCR) and cinnamyl alcohol dehydrogenase (CAD) to give rise to the three cinnamyl alcohols or monolignols: *p*-coumaric, coniferyl and sinapyl alcohol, which are the basic building blocks for lignin. The monolignols are thought to be exported to the cell wall (as glucosylated derivatives) where they undergo peroxidase- or laccase-catalysed polymerization (Figure 1).

Softwood and hardwood lignins

The proportions of each monolignol incorporated into lignin vary in different types of plant. Gymnosperm lignin consists predominantly of coniferyl alcohol units, yielding a guaiacyl lignin [5]. Angiosperm lignin contains both coniferyl and sinapyl alcohols and is known as guaiacyl-syringyl lignin. These differences in composition affect lignin structure and can influence lignin removal during pulping [6]. For example, guaiacyl-syringyl lignin is more easily hydrolysed than lignin made from guaiacyl units alone, making hardwoods easier to pulp than conifer softwoods. This is because the extra methoxy group at C-5 on the phenyl ring in sinapyl alcohol prevents condensed carbon–carbon linkages forming with the C-5 of another monomer during lignin polymerization. Guaiacyl lignin,

Figure 1

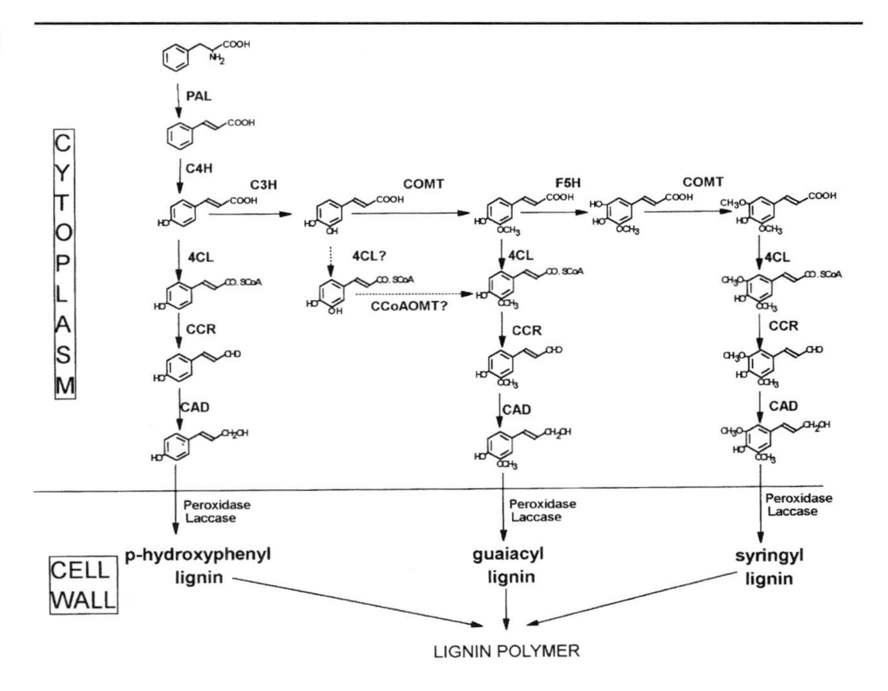

Lignin biosynthetic pathway

Although the pathway as far as the synthesis of the monolignols is known to occur intracellularly, the exact locations of individual reactions are unknown and could occur in the cytoplasm or be compartmentalized within organelles.

derived from coniferyl alcohol, which has a free C-5 position, has a higher proportion of these non-hydrolysable 5-5′ linkages than guaiacyl-syringyl lignin.

Lignin removal by pulping and bleaching

Chemical pulping using the Kraft method is one of the most widely used pulping processes. It involves cooking wood chips in a digester in sodium sulphide and sodium hydroxide at high temperature and pressure to dissolve most of the lignin. In determining the optimum chemical load and cooking time, there is a trade-off between achieving maximum lignin removal and maximum pulp yield. At the end of the cook the pulp is discharged from the digester, washed and separated into fibre and 'black liquor' components. The 'black liquor' contains dissolved lignin and spent chemicals and is burnt in a recovery furnace to provide energy for the cooking process and to recover chemicals. The pulp itself is still highly coloured due to the presence of residual lignin that must be removed by bleaching if the pulp is to be used for the production of fine paper. Again the degree of bleaching has to be carefully controlled since the chemicals that remove lignin will also damage the integrity of the cellulose fibres and hence the final yield and strength of paper produced. Bleaching commonly involves the use of chlorine compounds, often chlorine gas, and this aspect of the paper-making process has become a target for environmental concern since chlorine compounds can be very reactive and damaging to the environment. The industry is currently going to great lengths to reduce the environmental impact of pulp bleaching, primarily by reducing the use of chlorine compounds, particularly elemental chlorine.

Gene targets for modifying lignin biosynthesis

If the content or composition of lignin produced by woody plants could be altered to make pulping easier there would be significant benefits both for the industry itself and for the environment. In recent years, gene sequences have been determined for many enzymes of the phenylpropanoid pathway, including some dedicated to the lignin branch-pathway, e.g. [7,8]. This has allowed genetic approaches to be taken to the manipulation of lignin quality. To date, most work has concentrated on depressing the activity of enzymes involved in lignin production. This has been achieved by introducing antisense or partial sense genes for these enzymes into transgenic plants. Although the mechanism whereby such transgenes exert their effect is still debatable, their efficacy as tools for imposing specific blocks on biochemical pathways has been amply demonstrated in a variety of plant systems [9]. So far, the most suitable targets for lignin manipulation have proven to be the enzymes that are most specific for the lignin branch-pathway. It is obvious that interfering with enzymes of general phenylpropanoid metabolism, such as PAL, would affect many biological functions and might be seriously deleterious to the plant. This was inadvertently demonstrated by Elkind and co-workers [10] who tried to overexpress a heterologous bean PAL gene in tobacco. Paradoxically many of the transgenic plants had lower than normal PAL activities, presumably through homology-based silencing of the endogenous PAL gene. As a result of the low PAL activity, the plants had reduced lignin deposition in stem xylem but also exhibited a number of unusual and aberrant phenotypes, including stunted growth, altered leaf shape and texture, reduced pollen viability and altered flower morphology and pigmentation. Clearly, manipulation of PAL, and probably also of other early enzymes in the pathway, has too many diverse effects to be useful as a route to lignin modification. Appropriate target enzymes should be as specific as possible for the lignin branch-pathway so that other critical plant functions are not disrupted.

For the past 6 years, we have been collaborating with a network of European scientists to identify and manipulate enzymes specific and critical to lignin deposition. Three such genes were initially targeted: *CAD*, *COMT* and *CCR*. The expression of all three enzymes has been successfully manipulated in transgenic plants by antisense or partial sense transgenes. Analysis of lignin from these plants suggests that different changes in lignin composition can be achieved by targeting specific enzymes.

Manipulation of CAD activity

CAD, the final enzyme of lignin monomer synthesis, was initially purified, cloned and manipulated in tobacco (see [11,5,12] respectively). Tobacco was chosen as a model woody plant since large numbers of transgenics can be routinely and easily produced in this species and mature tobacco stems have a lignin content close to that of many trees.

Tobacco was transformed with a 1 kb fragment of the *CAD* cDNA inserted in antisense orientation under the control of the cauliflower mosaic virus 35S promoter. Despite large reductions in CAD enzyme activity (up to 95%), antisense plants developed a normal vegetative and floral organ morphology and growth rate. The final height and fresh weight yield were also indistinguishable from those of controls. Antisense plants with less than 16–20% CAD activity showed a novel red colour in xylem tissues at certain developmental stages. This colour was reminiscent of that seen in *bmr* lignin mutants and was found to be integrally associated with lignin, suggesting that the lignin had a novel composition. The total lignin content of antisense and control plants was determined by two methods. Both Klason and acetyl bromide estimations demonstrated that antisense plants incorporate the same amount of lignin as control plants. However, treatment of antisense and control lignin with sodium hydroxide or thioglycolic acid demonstrated great differences between the two samples. More lignin could be extracted from antisense samples by either treatment. This suggests structural changes in the antisense lignin had made it more extractable.

The composition of lignin from antisense plants was investigated by temperature-resolved in-source pyrolysis-mass spectrometry. The pyrolysis-MS profile obtained for control xylem was characterized by high-intensity mass peaks for coniferyl and sinapyl alcohol, indicative of a normal angiosperm lignin. Xylem from antisense plants gave a different profile. The peaks corresponding to coniferyl and sinapyl alcohols decreased in amplitude while those for coniferyl and sinapyl aldehydes increased. This is consistent with a block in CAD activity where the aldehyde substrates of the CAD reaction are incorporated into lignin in place of the usual reaction products, the monolignol alcohols. The altered aldehyde/alcohol incorporation was most obvious for the syringyl lignin component.

To determine whether the modifications achieved in lignin from CAD antisense plants was of use to the paper-making industry, control and antisense tobacco plants were subjected to a simulated alkaline Kraft pulping. The kappa number of pulps is an indication of the amount of lignin remaining in the material after cooking and the amount of chlorine needed to bleach the pulp to high brightness. Pulp produced from CAD antisense plants had a lower kappa number than pulp produced from control plants, demonstrating that the antisense plant fibres had been more efficiently delignified. Antisense plant pulp needed less chlorine at the bleaching stage to produce high quality paper. Microscopic examination of the pulps showed less cell clumping in the antisense plant sample, with fibres more dispersed and less coloured by lignin than in the control plant samples.

These results suggest a direct application of this work to the improvement of plants grown for pulp and paper production. Modification of lignin in trees or other fibre crops by genetic manipulation of CAD activity may reduce the costs of paper production by both lowering the costs of lignin removal and improving the yield. Most importantly, the possibility of either reducing the amount of bleaching chemicals used, or of replacing conventional bleaching agents with less reactive ones, will alleviate the damaging impact that pulping has on the environment. Similar CAD antisense constructs to those used in tobacco have now been introduced into poplar [13], a hardwood species commonly grown and pulped in N. Europe. Molecular, biochemical and chemical analysis of these plants show very similar changes in CAD expression and lignin composition to those previously seen in tobacco. Field trials of these novel trees are currently being performed in France and the U.K.

Manipulation of COMT activity

Atanassova and co-workers [14] have recently reported severe reduction of COMT activity in transgenic tobacco by introducing antisense or partial sense *COMT* transgenes. Phenotypically normal plants with as little as 3% remaining COMT activity were obtained. Unlike CAD antisense plants, lignin in COMT-reduced plants had a normal colour but staining with Maule reagent, which is specific for free syringyl units, suggested large changes in lignin composition. Whereas sections from wild-type plants stained bright red, sections from transgenic plants with low levels of COMT activity stained orange or yellow. This pointed to a marked decrease in free syringyl units in the lignin of low-COMT plants. Analysis of the lignin by more sophisticated methods confirmed this observation. Thioacidolysis is a technique that identifies monomeric products recovered from guaiacyl and syringyl units involved in β-O-4 bonds [15]. Since both units are major constituents of angiosperm lignin, their molar ratio is a characteristic feature of lignin. Thioacidolysis of lignin from low-COMT plants showed a marked decrease in syringyl moieties released, while guaiacyl moieties were unchanged or even slightly increased. Most interestingly, unusual 5-OH guaiacyl units not normally seen in lignin were detected. Klason lignin determination showed no significant change in total lignin content in low-COMT plants.

Almost identical results were obtained by Van Doorsselaere and co-workers [16] who used a similar strategy to decrease COMT activity in poplar. Plants with 50% reduced activity showed only slight changes in lignin monomer composition by thioacidolysis, but plants where COMT was reduced by 95% showed a 6-fold decrease in the syringyl:guaiacyl ratio. This was due both to a reduction in the number of syringyl units and an increase in the number of guaiacyl units. Again, novel 5-OH guaiacyl units were detected in thioacidolysis products from very-low-COMT plants. These same plants also showed a slight alteration in the colour of lignin, which changed from white–yellow to pale rose. However, two independent procedures, Klason acid precipitation and acetyl bromide solubilization, failed to detect any significant change in total lignin content, even in the most severely COMT-reduced plant lines.

The results of Atanassova and co-workers, and Van Doorsselaere and co-workers, while consistent with each other, contradict work where heterologous sequences were used to manipulate COMT activity in tobacco. Dwivedi and co-workers [17] reported alterations in both the composition and quantity of lignin in tobacco plants expressing antisense RNA for aspen COMT, while Ni et al. [18] described reduced lignin content but no change in composition in tobacco transformed with antisense alfalfa COMT. The reason for the discrepancies between these results is unclear, but may be something to do with the heterologous nature of the transgene used in [17] and [18].

Whether plants with reduced activity of COMT will be useful to industry has yet to be conclusively determined. A decrease in the methoxy groups in wood would be a benefit since these give rise to polluting thiols during pulping. However, achieving this simply by reducing the proportion of syringyl units is likely to lead to a more condensed polymer that is difficult to hydrolyse during pulping. Reducing syringyl units and replacing them with 5-OH guaiacyl moieties lacking the methyl group at C-5, might maintain lignin solubility while reducing thiol production. The results of Atanassova and co-workers, and Van Doorsselaere and co-workers, suggest that this may one day be possible and that incorporation of 5-OH guaiacyl units into lignin can be promoted by COMT reduction. However, in both low-COMT tobacco and poplar plants, relatively small amounts of 5-OH guaiacyl units were detected by thioacidolysis, and these are unlikely to completely substitute for the spectacular reduction observed in syringyl units.

Role of caffeoyl-CoA O-methyltransferase (CCoAOMT)?

Given the bifunctional nature of COMT, which can methylate both caffeate and 5-hydroxyferulate [19], the finding that in low-COMT plants the amount of guaiacyl lignin deposited increases slightly is very surprising. It has recently been suggested that another enzyme (CCoAOMT) is also involved in the methylation reactions important to lignin biosynthesis and may methylate both caffeoyl-CoA and 5-hydroxyferuloyl-CoA [20]. A persuasive body of circumstantial evidence has been presented to support a role for CCoAOMT in lignification [20,21] but only genetic experiments can confirm this. A role for CCoAOMT in the methylation of 5-hydroxyferuloyl-CoA is hard to reconcile with the huge reduction in syringyl lignin achieved by down-regulation of COMT but a role in the methylation of caffeoyl-CoA is very plausible.

Manipulation of CCR activity

Work is proceeding in two laboratories to determine the effects of reducing CCR activity in tobacco. A cDNA encoding this enzyme was isolated recently from *Eucalyptus* by scientists at the University Paul Sabatier in Toulouse [22], and was used as a probe to isolate the homologous gene from tobacco in our laboratory at Zeneca Plant Sciences. Our laboratories are now collaborating to produce and analyse tobacco plants expressing *CCR* antisense and partial sense genes. Plants with large reductions in CCR activity have been obtained and some show the altered phenotype of a brown colour in xylem tissue. Histochemical staining suggests that lignin composition is altered in these lines. Further analysis is currently being performed to determine exactly how the lignin composition and/or content have been altered.

Other targets for manipulation

Polymerization of lignin

Other workers have concentrated on manipulating the enzymes thought to be responsible for the polymerization of monolignol precursors into lignin. Controversy still exists about the identity of such enzymes, with both peroxidases and laccases being proposed as candidates.

Lagrimini and co-workers [23] overproduced the major anionic peroxidase of tobacco by introducing a chimaeric peroxidase gene under the control of the constitutively expressed CaMV 35S promoter. The resulting plants, with 10-fold increased peroxidase activity, had higher than usual levels of lignin in normal and wounded pith tissue. No examination of lignin levels in other tissues was reported. Transformed plants also had the unique phenotype of chronic severe wilting, through loss of turgor in leaves, which was initiated at time of flowering. However, plants expressing similar antisense constructs and with reduced peroxidase activity showed no change in lignin content or composition [24,25]. Additional experiments are necessary to determine whether peroxidases play a critical role in lignin polymerization, and whether they might provide suitable targets for lignin manipulation.

The first laccase genes were identified and cloned recently from *Acer pseudoplatanus* [26] and tobacco [27]. Overexpression and down-regulation experiments are currently underway in a number of laboratories to elucidate the role of these enzymes in lignification.

Syringyl lignin content

One frequently suggested objective for genetic engineering to improve pulping would be to modify the lignin in softwoods so that it would be more readily hydrolysed [28]. Although fibres from softwoods can be difficult to pulp, they are valued for their length and strength. Improving the ease of lignin removal might be achieved by introducing the hardwood genes that produce syringyl lignin into softwoods. It is as yet unclear how many genes would be required but it is possible that a small number might suffice [28]. Even for the improvement of hardwoods, increasing syringyl lignin, without increasing total lignin content, would be of benefit. However, none of the genetic manipulation experiments reported to date has been able to achieve an increase in syringyl groups. For example, increasing COMT activity by overexpressing the *COMT* gene in tobacco had no significant effect on lignin composition [14], indicating that O-methylation is not a rate-limiting step in syringyl lignin biosynthesis.

F5H, the enzyme preceding COMT in the conversion of ferulate into sinapate, is also an obvious target for manipulation. However, no gene sequences encoding this enzyme had been described until very recently when Meyer and co-workers reported the first cloning of a F5H gene from *Arabidopsis*, by T-DNA tagging [29]. *Arabidopsis fah1* mutants, defective in the F5H gene, do not make syringyl lignin [30], demonstrating an important role for F5H in determining lignin monomer composition. Meyer and co-workers suggest that F5H may catalyse a rate-limiting step in syringyl lignin biosynthesis. Their further work has supported this by demonstrating that overexpression of *Arabidopsis* F5H from a constitutive promoter leads to an increase in syringyl lignin content and abolishes the tissue specificity of syringyl deposition that is normally restricted to sclerified parenchyma ([31] and K. Meyer, personal communication). Whether introduction of F5H into conifers, or its overexpression in woody angiosperms, will be sufficient to increase the proportion of syringyl lignin and improve pulping will no doubt be the subject of much future research.

Conclusions

The work reviewed here highlights the opportunities that exist to improve plant fibres by genetic manipulation of lignin deposition. Lignin composition, and probably also lignin content, can be altered by down-regulating appropriate enzymes on the biosynthetic pathway. The exact changes in lignin achieved depend on the identity of the enzyme targeted, suggesting that it may be possible to tailor lignin composition for particular industrial processes. At least in the case of reduction of CAD activity, the resulting novel lignin has already been show to be more easily hydrolysed during pulping, offering an immediate opportunity for the improvement of trees grown for paper production.

The full potential for lignin modification needs to be explored by further experimentation. The results of antisense or overexpression experiments are often difficult to predict and the utility of novel lignins will certainly have to be assessed empirically. New experiments will include reducing or increasing the activities of combinations of genes that have already been manipulated singly. Attention will also focus on those genes for which results of genetic manipulation experiments have not been reported, as well as on identifying and manipulating genes for which clones are not yet available.

Clarifying the current uncertainty that exists over the identity of the enzymes catalysing particular steps of the lignin biosynthetic pathway is also an important objective. The confusion over the roles of COMT and CCoAOMT has already been mentioned but it is possible that alternative routes might also exist at other points in the lignin biosynthetic pathway [32]. Doubts still remain about the degree to which the pathway follows the same sequence in different plant species or even in different cell types [32]. Genetic manipulation experiments like those described here highlight this problem but are also the best way to address and clarify these uncertainties. Such work, therefore, not only presents possibilities for manipulating lignin to useful ends but also addresses basic scientific questions about a fundamental biochemical process unique to plants.

References

1. Sarkanen, K.V. and Ludwig, C.H. (1971) in Lignins: Occurrence, Formation, Structure and Reactions (Sarkanen, K.V. and Ludwig, C.H., eds.), pp.1–18, Wiley-Interscience, New York
2. Campbell, M.M. and Sederoff, R.R. (1996) Plant Physiol. **110**, 3–13
3. Cherney, J.H., Cherney, D.J.R., Akin, D.E. and Axtell, J.D. (1991) Adv. Agron. **46**, 157–198
4. Dixon, R.A. and Paiva, N.L. (1995) Plant Cell **7**, 1085–1097
5. Higuchi, T. (1985) in Biosynthesis and degradation of wood components (Higuchi, T., ed.), pp. 141–160, Academic Press, Orlando, FL
6. Chiang, V.L., Puumala, R.J. and Takeuchi, H. (1988) Tappi **71**, 173–176
7. Knight, M.E., Halpin, C. and Schuch, W. (1992) Plant Mol. Biol. **19**, 793–801
8. Bugos, R.C., Chiang, V.L.C. and Campbell, W.H. (1991) Plant Mol. Biol. **17**, 1203–1215
9. Kooter, J.M. and Mol, J.N.M. (1993) Curr. Opin. Biotechnol. **4**, 166–171
10. Elkind, Y., Edwards, R., Mavandad, M., Hendrick, S.A., Dixon, R.A. and Lamb, C.J. (1990) Proc. Natl. Acad. Sci. U.S.A. **87**, 9057–9061
11. Halpin, C., Knight, M.E., Grima-Pettenati, J., Goeffner, D., Boudet, A.M. and Schuch, W. (1992) Plant Physiol. **98**, 12–16
12. Halpin, C., Knight, M.E., Foxon, G.A., Campbell, M.M., Boudet, A.M. Boon, J.J., Chabbert, B., Tollier, M.-T. and Schuch, W. (1994) Plant J. **6**, 339–350
13. Baucher, M., Chabbert, B.G., Pilate, G., Van Doorsselaere, J., Tollier, M.-T., Petit-Conil, M., Cornu, D., Monties, B., Van Montagu, M., Inze, M., Jouanin, L. and Boerjan, W. (1996) Plant Physiol. **112**, 1479–1490
14. Atanassova, R., Favet, N., Martz, F., Chabbert, B., Tollier, M.-T., Monties, B., Fritig, B. and Legrand, M. (1995) Plant J. **8**, 465–477
15. Lapierre, C., Monties, B. and Rolando, C. (1986) Holzforschung **40**, 113–119
16. Van Doorsselaere, J., Baucher, M., Chognot, E., Chabbert, B., Tolliere, M.-T., Petit-Conil, M., Leple, J.-C., Pilat, G., Cornu, D., Monties, B., Van Montagu, M., Inze, D., Boerjan, W. and Jouanin, L. (1995) Plant J. **8**, 855–864
17. Dwivedi, U.N., Campbell, W.H., Yu, J., Datla, R.S.S., Bugos, R.C., Chiang, V.L. and Podila, G.K. (1994) Plant Mol. Biol. **26**, 61–71
18. Ni, W., Paiva, N.L. and Dixon, R.A. (1994) Transgenic Res. **3**, 120–126
19. Bugos, R.C., Chiang, V.L.C. and Campbell, W.H. (1992) Phytochemistry **31**, 1495–1498
20. Ye, Z.-H., Kneusel, R.E., Matern, U. and Varner, J.E. (1994) Plant Cell **6**, 1427–1439
21. Ye, Z.-H. and Varner, J.E. (1995) Plant Physiol. **108**, 459–467
22. Boudet, A.M., Lapierre, C. and Grima-Pettenati, J. (1995) New Phytol. **129**, 203–236
23. Lagrimini, L.M. (1991) Plant Physiol. **96**, 577–583
24. Chabbert, B., Monties, B., Liu, Y.T. and Lagrimini, M. (1992) in Proceedings 5th International Conference on Biotechnology in Pulp and Paper Industry (Kuwahara, M. and Shimada, M., eds.), pp. 481–485, Uni Publishers Co., Tokyo
25. Lagrimini, L.M., Gingas, V., Finger, F., Rothstein, S. and Liu, T.T.Y. (1997) Plant Physiol. **114**, 1187–1196
26. LaFayette, P.R., Eriksson, K.-E.L. and Dean, J.F.D. (1995) Plant Physiol. **107**, 667–668
27. Keifer-Meyer, M.-C., Gomord, V., O'Connell, A., Halpin, C. and Faye, L. (1996) Gene **178**, 205–207
28. Whetten, R. and Sederoff, R. (1991) Forest Ecol. Management **43**, 301–316
29. Meyer, K., Cusumano, J.C., Somerville, C. and Chapple, C.C.S. (1996) Proc. Natl. Acad. Sci. U.S.A. **93**, 6869–6874
30. Chapple, C.C.S., Vogt, T., Ellis, B.E. and Somerville, C.R. (1992) Plant Cell **4**, 1413–1424
31. Meyer, K., Cusumano, J.C., Ruegger, M.O. and Chapple, C.C.S. (1996) Plant Physiol. **111**(Suppl.), 143
32. Whetten, R. and Sederoff, R. (1995) Plant Cell **7**, 1001–1013

Engineering canola vegetable oil for food and industrial uses

Jean C. Kridl
Calgene, Inc., 1920 Fifth Street, Davis, CA 95616, U.S.A.

Plant oils: uses, biosynthesis and engineering

Plant oils have numerous food and industrial end uses. These include edible oils, processed ingredients for the food industry, feedstocks for chemical processes, including formulations of paints, lubricants and plastics, and medical applications. The development of widespread applications of plant oils in such diverse industries is primarily a result of the availability and the variety of oils of different compositions made by plants and the functionality of these oils. For example, the rapeseed plant (*Brassica* sp.), which has oil traditionally high in the 20-carbon, monounsaturated (20:1) fatty acid erucic acid, used in industrial applications, has been altered by plant breeding to produce an edible oil; canola, which is high in 18-carbon mono- and polyunsaturated fatty acids [oleic (18:1), linoleic (18:2) and linolenic (18:3)]. The fatty acid composition of these two oils is dramatically different and, as a result, so are their functionalities and product applications.

The oil biosynthetic and storage pathways have been heavily investigated at the biochemical and more recently at the molecular level (reviewed in [1]; Figure 1). It is now commonly accepted that plants synthesize 18-carbon fatty acids through a common pathway located in the plastid that elongates acetyl-CoA in two-carbon increments using malonyl-acyl carrier protein (ACP) as the carbon donor. The first desaturation step occurs inside the plastid and the fatty acids, palmitate (16:0), stearate (18:0) and oleate (18:1), are made available by the action of acyl-ACP thioesterase for further elongation, desaturation and incorporation into membrane and storage lipids in the cytoplasm on endoplasmic reticulum membranes. Unusual fatty acids are made and stored in seed oils as triacylglycerols (TAGs) through modifications of this basic pathway. For instance, in California bay laurel, a specialized thioesterase with specificity for 12:0-ACP is expressed during seed development [2,3] and is at least in part responsible for the accumulation of high levels of laurate in the bay seeds.

Armed with a basic understanding of the biochemical pathways controlling the biosynthesis of fatty acids, it was suggested early on that oil biosynthesis could be modified by genetic engineering to produce oils of any composition for directed applications [4]. In theory, this genetic engineering approach required cloning of the genes controlling certain steps of biosynthesis, regulating the genes for proper expression in the seeds of plants and a transformation and regeneration system for the oil-seed of choice. Essentially this approach has been demonstrated in the temperate crops canola and soybean

Figure 1

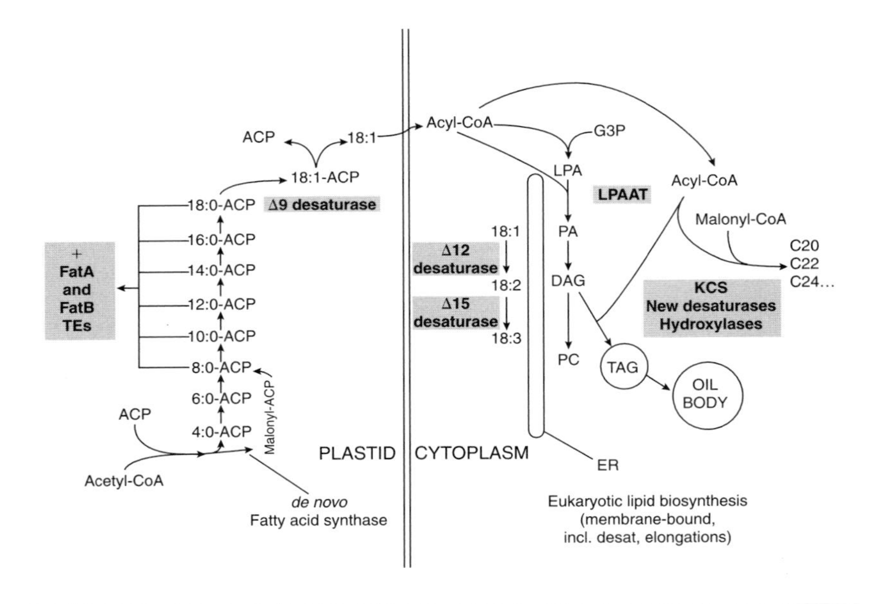

Schematic representation of seed oil biosynthesis

Fatty acid biosynthesis takes place inside the plastid, while modifications of fatty acids and incorporation into membranes and storage triacylglycerol occurs in the cytoplasm. Enzymes highlighted have been manipulated by engineering (Table 1). Abbreviations; KCS, β-ketoacyl-CoA synthetase; LPA, lysophosphatidic acid; LPAAT, lysophosphatidic acid acyltransferase; PA, phosphatidic acid; DAG, diacylglycerol; PC, phosphatidylcholine.

(Table 1) and should shortly be accomplished in more transformation recalcitrant species such as sunflower and maize. However, commercialization of these oils is a reality in only one case so far and in others it is requiring additional basic research effort to understand the effect of oil modification on the seed physiology.

This paper will focus on two products being developed at Calgene for the industrial and edible oils markets. One is high-laurate oil and the other is high-stearate oil. For both products, the strategy was to increase the level of a particular chain-length saturated fatty acid in the seed oil. In the case of laurate this involved adding a single gene to intercept the chain elongation pathway at 12 carbons rather than the normal 18 carbons. In the case of stearate, one strategy was to turn off the gene encoding the first desaturation step, thereby increasing the pool of 18:0 available for incorporation into seed oil. As will be described in some detail below, both of these steps were achieved in a relatively short time-frame and oil modifications were observed from both strategies, but the path to commercialization for the two oils has been quite different.

Table 1

Fatty acid modification	Name	Gene	Transgenic plant oil	Reference
↑8:0	Capric	FatB TE	Canola	[17,18]
↑ 10:0	Caprylic	FatB TE	Canola	[17,18]
↑ 12:0	Laurate	FatB TE	*Arabidopsis*	[5]
			Canola	[6]
↑ 14:0	Myristate	FatB TE	Canola	[18], Dehesh, K. and Voelker, T.A.†
↑ 16:0	Palmitate	FatB TE	Canola	[16]; [18], Hitz, W.D.†
			Soybean	Kinney, A.J.†
			Arabidopsis	[15]
		FatA TE	Soybean	Kinney, A.J.†
↓16:0	Palmitate	FatB TE	Soybean	Kinney, A.J.†
		KAS	Canola	Thompson, G. and Lassner, M.W.†
↑ 18:0	Stearate	Δ^9-Desaturase	Canola	[13]
		FatA TE	Canola	This paper, Hitz, W.D.†
		Δ^9-Desaturase/ FatA TE	Canola	Kinney, A.J.†
↓ 18:0	Stearate	Δ^9-Desaturase	Canola	J.C. Kridl and G. Thompson‡
↑ 18:1Δ^9	Oleic	Δ^{12}-Desaturase	Soybean	[19]
			Canola	[19]

Table 1 (contd.)

Fatty acid modification	Name	Gene	Transgenic plant oil	Reference
↑ $18:1\Delta^{6}$	Petroselenic	Δ^{4}-Desaturase	Sunflower	[20]
↑ $18:1\Delta^{12OH}$	Ricinoleic	Δ^{12}-Hydroxylase	Tobacco	[21]
↑ $18:2\Delta^{9,12}$	Linoleic	Δ^{15}-Desaturase	Canola	[19]
↓ $18:3\Delta^{9,12,15}$	α-Linolenic	Δ^{15}-Desaturase	Canola	[19]
↑ $18:3\Delta^{9,12,15}$	α-Linolenic	Δ^{15}-Desaturase	*Arabidopsis*	[22]
↑ $18:3\Delta^{6,9,12}$	γ-Linolenic	Δ^{6}-Desaturase	Tobacco	[23]
			Arabidopsis	[23]
↑ 22:1	Erucic	LPAAT	HEAR	[24]
↑ 24:1	Nervonic	KCS	HEAR	Lassner, M.A.†
↑ 22:1OH	Alcohol/wax	Acyl-CoA reductase	HEAR	[25]

**TE = thioesterase.*

†Personal communication.

‡Unpublished work.

Modified seed oils in transgenic plants

Laurate canola

Laurate canola is currently available for sale under the brand name Laurical™. The oil was developed by a genetic engineering strategy in which a specialized acyl-ACP thioesterase gene from the California bay tree was expressed in the seeds of canola. California bay accumulates ~60% of its fatty acids as laurate and a bay thioesterase, which showed specificity for 12:0-ACP as a substrate [2] was isolated using protein purification techniques [3]. The protein was sequenced, gene probes generated and the complete cDNA clone was isolated from a developing seed library [5]. This gene is limited in its expression to bay laurel seeds and is at least in part responsible for the accumulation of high levels of laurate in the seed oil. Seed-specific expression of this gene from a napin promoter in *Arabidopsis* [5] and canola seeds [6] showed that the enzyme could interact effectively with the fatty acid biosynthetic enzymes to produce high levels of laurate in seed oil. The highest laurate levels in pooled canola seed of the initial set of transgenic plants was 40 mol% laurate. This plant contained as many as 15 copies of the transgene with greater than four genomic locations [6]. Through crossing, selection and plant breeding, some of these transgenes have been maintained and a stable seed line has been produced that has ~46 mol% (38 wt%) laurate. It is oil from this line that is currently being produced for the Laurical product. Lines with higher levels of laurate (up to 58 mol%) are currently in field testing.

It is interesting to note that the initial canola plant phenotype was obtained in 1991–1992. The seed yield of the initial transformant chosen for increase was only 50% of controls, but by 1993 the yield had been increased to 70% of controls through selection and breeding. The detrimental attributes of the initial transformant have subsequently been shown not to be due to the laurate genes since current selections from lines crossed to canola varieties have equal yields to check varieties [7]. The oil yield in the high-laurate lines is normal and laurate-containing TAGs are utilized well by germinating seeds. Some 1.2×10^6 lb (1 lb ≈ 0.454 kg) of Laurical were produced from seed grown in Michigan in the summer of 1995, and a crop in North Dakota yielded approx. 3×10^6 lbs of oil in November 1996 that is being sold to selected segments of the food industry.

Laurate applications

Laurate canola is unique in its TAG structure as well as its composition of laurate. The total amount of laurate in Laurical approaches that found in coconut oil and palm kernel oil [8], the current sources of most laurate used in food and industrial applications. However, the TAGs of laurate canola have predominantly the structure L-U-L where laurate (L) is in the 1- and 3-position of the glycerol backbone and the unsaturated (U) fatty acids 18:1, 18:2 and 18:3 occupy the sn-2 position. This is in contrast to coconut oil where a greater proportion of the TAGs are trilaurin (L-L-L). The TAG structure of laurate canola may offer unique functionality to the oil. Applications testing is currently underway to determine the exact properties of the oil in products that currently use lauric oils and where high melting temperatures are desirable. Some suggested applications are confectionery coatings, coffee whiteners and other simulated dairy products, icings and bakery fillings as well as margarines and spreads. Preliminary results using hydrogenated laurate canola in confectionery coating formulations show that Laurical gives significant improvement in flavour release as compared with currently used palm kernel oil fractions, is highly compatible with cocoa butter and confers increased stability to chocolate 'bloom' [9]. The challenge for the future will be applications development to incorporate this novel oil into other food systems.

Industrial applications and the lysophosphatidic acid acyl-transferase (LPAAT) gene

There are a number of industrial applications for lauric oils, the primary one being as an ingredient in soaps and detergents. However, the economics of this market demand that much larger amounts of lauric acid be produced in canola to be of commercial interest. As stated above, laurate is essentially excluded from the sn-2 position of the TAGs in canola and this limits the total amount of laurate that can be accumulated. It was suggested some time ago that it was the specificity of the enzyme LPAAT, which puts the fatty acid in the sn-2 position, that would limit accumulation of unusual fatty acids, both long and short chains, in temperate crop plants [10]. The LPAAT from canola clearly prefers 18:1-CoA as a substrate. The LPAAT enzyme from coconut was shown to have a specificity for medium-chain-length substrates [11] and therefore became an engineering target for a second laurate product.

LPAAT was solubilized [11], partially purified and sequenced from developing coconut endosperm [12]. The gene was cloned using degenerate probes to peptide sequences and its identity as an LPAAT was verified by expression in *E. coli*. The enzyme showed distinct specificity for the medium-chain substrates 10:0-, 12:0- and 14:0-CoA with the majority of the activity on 12:0-CoA. This gene has since been expressed under seed-specific control in canola (D.S. Knutzon and H.M. Davies, unpublished work). LPAAT activity was detected in seed extracts from a number of different transgenic plants. The plants with the highest activities were crossed with high-laurate plants. The seeds from this cross were analysed for laurate composition and for TAG structure. Significant accumulation of laurate was seen in all three positions of the TAG, unlike in the plants with bay thioesterase alone. The overall level of laurate does not seem to be increased in these plants, but both the LPAAT and bay thioesterase genes are segregating in the seed population, therefore further breeding work may affect the overall laurate accumulation. A number of years of plant breeding and field evaluation will also be necessary to determine if higher laurate levels or the trilaurin TAG structure affect the physiology of the plants or seeds in any way.

Stearate canola

Antisense strategy

The first modified-oil's phenotype reported from genetic engineering of the oil's biosynthetic pathway was a high-stearate canola [13]. The relative activities of the enzymes stearoyl-ACP desaturase, oleoyl-ACP thioesterase and β-ketoacyl-ACP synthase determine the amount of stearate that is naturally found in canola oil (Figure 1). Normal levels of stearate are quite low, ~2 wt%. Down-regulation of the gene encoding stearoyl-ACP desaturase via antisense gene expression allowed accumulation of high levels of stearate in seed oil [13]. The best initial transgenic plant had pooled seed stearate levels around 15% with single seeds as high as 40%. The plant, designated A3, had at least three genetic loci containing antisense desaturase genes under control of the seed-specific promoter, napin. As with the high-laurate oils, the saturated fatty acid stearate does not tend to go into the second position of the TAG and therefore high-stearate oil has a unique structure. A high-stearate oil of this structure has considerable commercial interest in the production of margarines and shortenings for baking applications as well as for confectionery applications as a cocoa butter substitute. Products made from a high-stearate canola oil would have no *trans* fatty acids and might be less cholesterolaemic than palm or hydrogenated oils.

The high-stearate phenotype was reported in 1992 [13] at about the same time as the laurate plants were obtained [5]. Extensive plant breeding has been undertaken with the initial transgenic plant, A3, as well as with new transgenic lines to develop a commercial high-stearate product. Plants have been selected that contain one of the original stearate loci from A3 and that have levels of stearate >30% in different genetic backgrounds. The selection, greenhouse and field growth of these high-stearate plants has led to a number of observations concerning their agronomic properties and is leading to new lines of investigations concerning fatty acid utilization in these plants. Initial observations indicated that the stearate levels could vary widely between seeds in any seed population [as much as 2% (background levels) to 40% stearate]. This was true even in plants homozygous for the introduced antisense desaturase gene. In addition, a wide variation in stearate levels was observed between sibling plants as well as the seed to seed variation. The variation was not due simply to background effects of other genes on the transgenes because the same phenomenon was observed in dihaploid plants obtained from microspore culture that are (in theory) genetically homozygous at all loci. Selections of half-seeds containing high levels of stearate from dihaploid plants and growth of the individuals again led to seed populations with a range of stearate concentrations. Stearate concentrations ranging from higher to much lower than the original half-seed selection have been observed. The variability observed suggests an environmental or physiological influence on the local accumulation of stearate. It has also been suggested that the observed variation arises from the poorly understood antisense mechanism by which the stearoyl-ACP desaturase gene has been turned down. However, evidence for gene silencing via methylation or interaction of the transgenes with endogenous genes has been negative to date (Mike Lassner, personal communication).

Another type of instability has been observed with the high-stearate plants over field generations. Unless selective pressure is maintained the stearate levels drop significantly from generation to generation in the field. For example, a third generation selection of the original transformant was grown during the summer of 1992 in field plots. The seed was harvested and then planted, without selection, in southern California in the winter of 1992. That seed was harvested and planted, again without selection, in Michigan in the summer of 1993. Seed from each successive generation was planted in plots in Michigan in the summer of 1994 to compare their performance under identical environmental conditions. The results indicated that the first non-selected generation seed had an average stearate content of 26.8% stearate, the second generation had 20.2% stearate and the third had 12.5% stearate. A decline of approximately 5% per generation was not unusual in similar experiments. These data suggested that there could be a physiological or environmental influence on the viability or vigour of high-stearate material.

Initial observations on seed germination indicated that seed populations with stearate levels between 20 and 30% and with total saturates slightly above 30% had normal seed yield and germination rates. However, as selection and breeding continued and populations were developed that gave stearate levels >30% with total saturates greater than 40%, a slight, but significant, decline in germination has been observed and the seedlings are less vigorous. Cotyledons of seeds with >30% stearate often have yellow spots and in severe cases necrotic lesions on the inner and outer cotyledons. The necrotic areas show higher levels of stearate than the surrounding normal tissue. It has been suggested that stearate may be accumulating in membrane as well as storage lipids in the high-stearate plants and this may cause the impaired germination and seedling vigour. Investigations of stearate levels in different lipid classes in mature seeds and during germination show that stearate has accumulated to significant levels in membrane-associated lipid classes (Figure 2) and that structured TAGs of the S-U-S type are not broken down as readily as less saturated TAGs during germination (Greg Thompson, unpublished work). In contrast, laurate is also present in the membrane lipids in high-laurate seeds at maturity (though at only 20% of the laurate found in the neutral lipids; Toni Voelker, unpublished work), but laurate TAGs appear to be readily utilized. Future investigations will focus on lipid classes in developing embryos during the time of stearate accumulation to assess the impact of stearate on seed and seedling physiology.

Thioesterase strategy

The observations described above have raised questions as to whether it is the mechanism by which stearate was produced in these plants or if it is accumulation of stearate itself or the level of stearate that leads to the stearate instability and the poor seed performance. We have recently used a different approach to producing high levels of stearate in storage TAGs of canola. This strategy is based on an acyl-ACP thioesterase. Investigations of the thioesterase gene family of a number of plant species has led to the classification of thioesterases into two groups based on nucleic acid sequence, namely FatA and FatB [14]. The members of the FatA class are all oleoyl-ACP thioesterases while the FatB class has specificity for the saturated substrates. Some FatB members have unique substrate specificity's for a particular chain-length fatty acid, such as the bay thioesterase for C12:0-ACP [5], while others, that appear to be ubiquitous in plants, have a broader saturated chain-length specificity centred on 16:0-ACP, such as those described for *Arabidopsis* [15] and *Cuphea* [16].

There are a few species that naturally accumulate high levels of stearate in their seed oils. Some of these are cocoa (*Theobroma cacao*, 35% 18:0), shea

Complex lipid biosynthesis and its manipulation in plants

Antoni R. Slabas* and Sherrie L. Sanda
Lipid Molecular Biology Group, Department of Biological Sciences, University of Durham, South Road, Durham DH1 3LE, U.K.

The central importance of lipids in metabolism has been recognized for some time [1,2]. If one lists the roles of lipids in cells it is quite formidable and new roles are increasingly being found as investigations into a diverse array of biological problems are being carried out [3,4]. The main function attributed to lipids is that of structure in the form of membranes. Membranes separate the cell into compartments with different metabolic roles. They act as barriers to metabolites and have inserted in them specialized transporters for ions and water (aquaporins) [5]. In addition, the lipid components are believed to play a crucial role in maintaining the fluidity of membranes during temperature decreases to prevent the reduction of essential cell functions to a level that is unacceptably low for cell viability [6,7]. Membranes, such as the plasma membrane, the endoplasmic reticulum and those present in mitochondria and chloroplasts all have different lipid compositions [8]. The cell must therefore have mechanisms to maintain the differences despite the role of lipids in membrane trafficking, which could be expected to lead to randomization of lipid composition [9]. There is clearly some biological basis for these differences in organelle lipid composition but at the moment this is not known.

In addition to a structural role, lipids are also involved in many other functions in cells. They act as key intermediates in cell signalling pathways (e.g. phosphatidylinositol, jasmonic acid) [10,11]. Lipids also provide the environment for specialized functions such as electron transport and desaturation and for acylation reactions to take place. They are important components in the post-translational modification of proteins, which permits their specific insertion into membranes [12]. Additionally, they play a role in sensing changes in the environment [13] and function as an energy store in the form of triacylglycerols.

Research into lipid biosynthesis has expanded dramatically over the last few years and has been considerably influenced by the potential of genetically engineering plants. There is considerable scope for engineering new oils for industrial use as well as trying to understand the genetic and biochemical basis for diversity of lipid types [14,15]. This has been fuelled by both industrial interests and the development of new techniques that have aided the elucidation of metabolic pathways and the isolation of genes that code for key components. Much of the interest in this area has been centred around the components responsible for the biosynthesis of fatty acids. This has included research into the following: (1) acetyl-CoA carboxylase, which provides the substrate for *de novo*

* To whom correspondence should be addressed.

fatty acid biosynthesis [16,17]; (2) fatty acid synthetase, which synthesizes fatty acids of chain length from C-4 to C 18:0 (reviewed in [18]); (3) thioesterases that are responsible for chain termination [19,20]; and (4) desaturases responsible for introduction of unsaturation into lipids [21–23]. Considerable advances have been made in understanding the enzymology, including the determination of the three-dimensional structures of at least two enzymes of plant lipid biosynthesis; enoyl ACP reductase [24], and Δ^9 desaturase [25]. Our studies on enoyl reductase, which have been conducted in collaboration with David Rice and John Rafferty at Sheffield, U.K. and Toon Stuitje at Amsterdam, have implications for designing inhibitors of the enzyme aimed at the treatment of tuberculosis [26].

There is little doubt that structural studies will become increasingly important over the next few years as the potential for altering substrate specificity based on knowledge of three-dimensional structures becomes apparent. We currently have one such research programme, aimed at converting an enoyl ACP reductase into a β-ketoreductase that will potentially result in completely different redox chemistry.

Some time ago it became obvious that most of the central enzymes of lipid synthesis would soon be cloned, especially with the commencement of large international efforts aimed at random (expressed sequence tag) sequencing, e.g. in *Arabidopsis* and *Ricinus* [27,28]. Genome size is no longer an important consideration as larger companies such as Pioneer Hi-Bred have realized the strategic importance of genome sequencing of crops that are central to their business concerns. With this work in progress it is necessary to stop and consider the scientific opportunities. Availability of all the DNA sequences will still not indicate function unless there is close homology to characterized sequences already in the databases. It was at this point that our research interests started to move to the synthesis of complex lipids, which had hitherto been a fairly neglected area. This article concentrates on some of the components that we have studied, with our collaborators, and the lessons that we have learned.

The biosynthesis of phosphatidylcholine and the cloning of plant genes in the choline phosphate pathway by complementation of yeast mutants

The pathway for the biosynthesis of phosphatidylcholine (PC) in plants is plastic. PC in plants can originate from cytidine diphosphate-choline (CDP-choline) (via the choline pathway) but there are also routes involving phosphatidylethanolamine (PE) and phosphatidylserine (PS) [29]. The pathway in *Saccharomyces* has been well-studied and the genes for all the component enzymes have been characterized [30]. The route via choline involves the three enzymes, choline kinase [EC 2.7.1.32; CKI], CTP:phosphocholine cytidyltransferase [EC 2.7.1.15; CCT], where CTP = cytidine triphosphate, and CDP-choline:diacylglycerol choline-phosphotransferase [EC 2.7.8.2; CPT]. CCT is apparently involved in the rate-limiting step in this pathway. The isolation of the cDNA for CCT from rape came about as a collaboration between Ikuo Nishida in Tokyo and our group in Durham [31]. Initial attempts were made to clone the cDNA using complementation of a yeast CCT mutant. This line of investigation failed to isolate a CCT clone in Durham but the key was provided by Ikuo on his return to Japan by engineering specific mutants for the complementation experiment – all the existing mutants were too leaky. There was also a major problem with the viability of mutants in the PS synthetase gene (CHO1). To overcome this a yeast strain was engineered to contain CHO1 under the control of the inducible galactose promoter. In this way the PS pathway could be switched on or off at will and the existing yeast CCT gene could then be disrupted using homologous recombination. The new mutant could then be used to clone the rape CCT cDNA by transforming it with a plant plasmid library in a yeast shuttle vector system and then switching off the PS pathway (Figure 1). In this way the cDNA for the CCT gene was cloned from rape. It should be noted that Ralph Dewey previously used complementation cloning in yeast to clone the plant CPT gene, which was the first member of the plant choline phosphate pathway for PS synthesis to be cloned [32].

The lesson learned from this work is that the original strategy may be correct but it is liable to require refinement, especially if entering into a new area of science. In our case the extra effort required to create additional mutants was fully justified.

Cloning of acyltransferase and attempts to clone malonyl-CoA:ACP transacylase (MCAT) by complementation of *E. coli* mutants

Before the yeast CCT study we decided to attempt complementation cloning as a way to obtain enzymes that had defied purification and were not amenable to

Figure 1

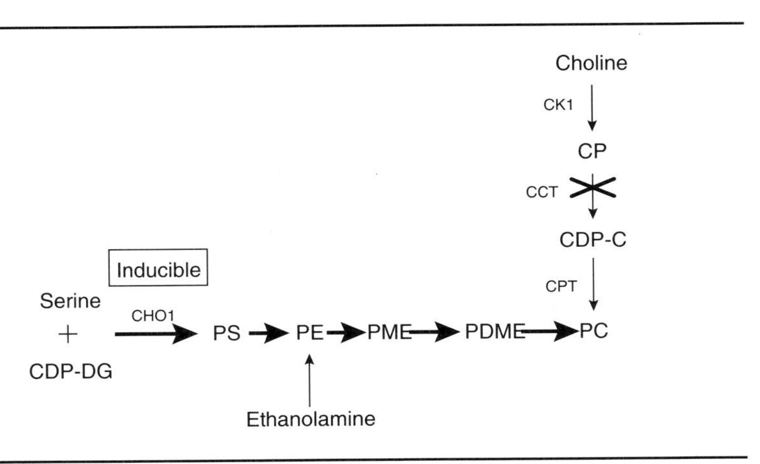

Biosynthetic pathway for PC in the yeast strain used to clone CCT

A yeast strain was engineered to contain a mutation in the CCT gene and carry a CHO1 gene under the control of an inducible galactose promoter. After transformation with a rape cDNA library into this strain the CHO1 gene was turned off and mutants complementing the PC pathway were selected. Thick arrows indicate the primary PC production pathway. Abbreviations not defined in the text: CP, choline phosphate; CDP-DG, cytidine diphosphate diacylglycerol; CDP-C, CDP-choline; PME, phosphatidylmonomethyl-ethanolamine; PDME, phosphatidyldimethylethanolamine.

classical biochemical approaches. The essential components and advantages of cloning plant genes by complementation of *E. coli* mutants are: (1) a mutant in *E. coli* that does not have a high reversion rate has to be available. Temperature-sensitive mutants are often desirable as they are easier to keep viable. (2) A plasmid cDNA library that contains the full-length cDNA of interest needs to be available from the plant source. In this way the plant protein can be synthesized in the bacteria and complement the mutation. Truncated cDNA would produce a truncated protein that may not be functional. (3) Neither the nucleotide sequence nor the amino acid sequence needs to be the same as that of the *E. coli* counterpart as the basis of selection is biochemical complementation. (4) Selection can be performed at very high plate densities of the *E. coli* as only the complemented *E. coli* will grow. (5) Complementation is only dependent on the biochemical product being made. It does not matter if it is located in a membrane or in the soluble phase of the cell, providing it is available to the enzymes that further metabolize it. (6) There could be an advantage in using complementation with membrane-bound enzymes as the protein may well be inserted into a membrane. This will allow direct assay of substrate specificity in microsomes and overcome the need for reconstitution in a lipid micelle.

We selected three targets to attempt cloning by complementation using known mutants available in *E. coli*: (1) membrane-bound 1-acyl-glycerol-3-phosphate-acyltransferase (2AT) [33]; (2) MCAT [34]; and (3) glycerol-3-phosphate-1-acyltransferase (1AT) [35].

The project was basically thought up over a cup of coffee and owed much to chance. Andy Tommey had been part of a group that had cloned maize dihydropicolinate synthetase, an essential component of lysine metabolism, by complementation cloning in an *E. coli* mutant [36] and he agreed to collaborate with us on our complementation efforts. The technique was novel and required full-length cDNA to work. Andy provided a maize plasmid library that had been used successfully to clone plant glutamine synthetase by complementation and contained full-length cDNA [37]. We focused our efforts on the membrane-bound 2AT, which proved to be the better choice of our three targets as discovered later when attempted complementation of the other two targets resulted in cloning of the wrong genes. Adrian Brown, a postdoctoral worker in our laboratory, carried out the task of complementing the 2AT mutation with the advice and assistance of Tommey. Fortunately, there were very few apparently successful transformants on the experimental plate (which had apparently complemented the mutant) compared with the control. Nevertheless only one complementation is required to clone the gene. Plasmid DNA was made from all the apparent transformants that complemented the 2AT mutation and used to retransform the mutant. Nearly all of the cells transformed with these cDNAs survived (Figure 2) and sequencing of the insert from one plasmid (pMAT) showed strong homology with the *E. coli* 2AT gene *plsC*, including several conserved domains [38]. One of the key elements in this success was the advice and help from Jack Coleman of Louisiana State University on how to grow the mutant that he provided. We were not so fortunate with complementation of the MCAT mutant. While complementation was achieved, when the cDNA was sequenced it showed the strongest homology to a G-type protein and the

Figure 2

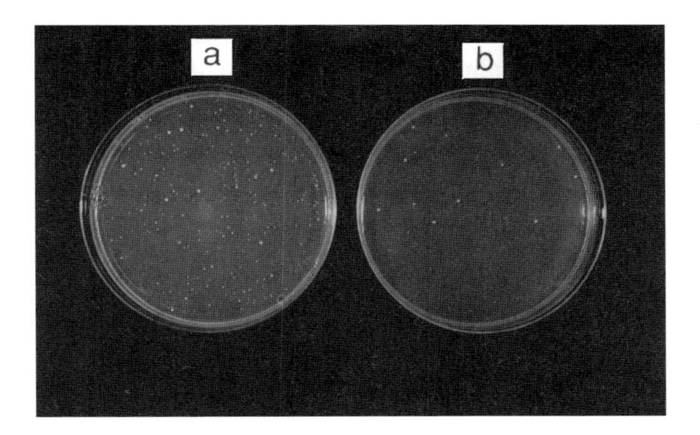

Cloning of the 2AT gene from rape via complementation of an *E. coli* mutant

The E. coli mutant JC201, which contains a mutation in the 2AT gene (plsC), was transformed with a rape cDNA library. (a) Complementation of the E. coli mutation by plasmids carrying rape DNA that successfully restored growth on the first round of transformation. (b) E. coli mutants transformed with plasmid not containing plant DNA.

biochemical basis of this complementation is still not understood [39]. Other workers have also tried to clone the plant 1AT by complementation of the *E. coli* mutant but isolated a glyceraldehyde-3-phosphate dehydrogenase cDNA [40]. The physiological reason for this particular complementation is now known: since the original mutant affected the K_m of the enzyme for glycerol 3-phosphate, anything that increased the pool size of glycerol 3-phosphate would complement.

The lesson learned from this study is that complementation cloning of *E. coli* mutants works but does not always identify the required cDNA. In particular, it is important to contact the person who has provided the mutant to obtain as much advice as possible.

Cloning of *Limnanthes*-2-AT with specificity for erucic acid

The desirability of being able to direct specific fatty acids into certain positions on the triacylglycerol backbone was recognized several years ago. Cao and co-workers [41] demonstrated that the acyltransferases of plants did not all have the same specificity and that microsomes from oil-seed rape would not incorporate erucic acid C22:1 (Δ^{13}) into position 2 whilst those from other species could incorporate erucic acid. This limits the upper yield of C22:1 in rape to 66%. As it was desirable to have a higher level of erucic acid in oil-seed rape, we decided to clone the 2-acyltransferase (2-AT) from *Limnanthes douglasii,* which was known to incorporate C22:1 into position 2. Clare Brough and Adrian Brown, in Durham, used two strategies to clone 2-ATs from *Limnanthes douglasii*: complementation cloning as used for the cloning of pMAT above, and the use of a heterologous pMAT probe. Two different types of 2-AT clones were isolated. The first approach resulted in the isolation of pLAT2 while the second resulted in the isolation of a different clone, pLAT1 [42]. The properties of these two clones are shown in Table 1.

Subsequent substrate specificity studies using 1-erucoyl-*sn*-gylcerol-3-phosphate (C22:1-LPA) demonstrated that pLAT2 would use C22:1-CoA as well as C18:1-CoA. Clare Brough therefore placed both of the pLAT constructs under the control of the napin promoter and Tina Barsby and her team at Nickerson Biocem (Cambridge) transformed them into oil-seed rape using the *Agrobacterium* method. So what should we look for in the transformants apart from the selectable marker? The answer was trierucin as this would be absent from the wild type, which could not incorporate erucic acid into position 2, but would be present in transformants if the 2-AT from *Limnanthes* was functional *in vivo*. It was as well that I had advised Nickerson to collaborate with Bill Christie (Scottish Crop Research Institute, Dundee) in some lipid analyses a few months previous as he was able to analyse the transformants using his HPLC-based system. While the trierucin level was not high (2.8% on the first run), this was confirmed by mass spectroscopy and, to our surprise, positional analysis showed that the level of C22:1 in position 2 was 28% in the pLAT2 transformant as opposed to 0% in the wild type [43]. This work is still ongoing.

The lesson learned from this work is to keep as many options open as possible when searching for a gene. In addition, seek assistance from experts in transformation: these will often be in industry. Analyses of substrate specificity for membrane-bound enzymes following complementation can be an excellent selection method, indicating the most promising clones. Get an analytical chemist on your side!

Towards the cloning of genes for complex lipid biosynthesis – a mutagenic approach

We have been very impressed by the advances made by Chris Somerville and John Browse who both used mutagenesis to elucidate metabolic pathways and isolate genes of lipid metabolism [44]. Such an approach could be used to isolate genes for complex lipid synthesis. The key to their success, apart from considerable foresight, was the ability to perform rapid analyses of the fatty acid composition of leaf material. They developed a rapid, high throughput GLC method that was

Table 1

pLAT1	pLAT2
Isolated by heterologous probing using pMAT1	Isolated by complementation cloning of JC201
mRNA in embryo, leaf and stem	mRNA in embryo only
1.5 kb cDNA, 377 aa, 42.7 kDa protein	1.1 kb cDNA, 281 aa, 31.6 kDa protein
66.8% identity to pMAT1	
21.5% identity to yeast SLC1	27.9% identity to *E. coli* PLS

Characteristics of the two *Limanthes* 2-AT clones

not very labour-intensive and allowed them to analyse a large number of samples. The situation is different with complex lipid analyses because great care has to be taken to prevent lipases breaking down the complex lipids, and the necessary analytical techniques are only available in a few very specialized laboratories. Chris Somerville was contacted for his advise on the feasibility of obtaining viable lipid mutants and the best way in which to proceed. The reply was a positive one and he exemplified the success of Christoph Benning in the isolation of a mutant in the biosynthesis of digalactosyldiacylglycerol (DGDG) from *Arabidopsis* [45]. Christoph Benning gave us advice on the problems we should consider in designing a lipid mutant screen. In addition, he provided us with mutagenized seed that he had prepared as it was likely that commercially available seed would have a low proportion of slow growing plants that might contain the very mutants we required. The analytical problems in the identification of a lipid mutant were solved by using the recently developed ternary gradient HPLC system of Bill Christie [46]. A lipid mutant screen was therefore developed that involves the extraction of lipids from the leaf tissue of individual mutagenized plants and subsequent detection of the lipid classes by HPLC analysis (Figure 3). This is an ongoing collaboration with Bill Christie and we are looking for mutants in which classes of individual lipids are altered or absent. At the moment the major lipid components have been identified in relation to the retention times of standards on the HPLC and further characterization of unidentified peaks is being performed

Figure 3

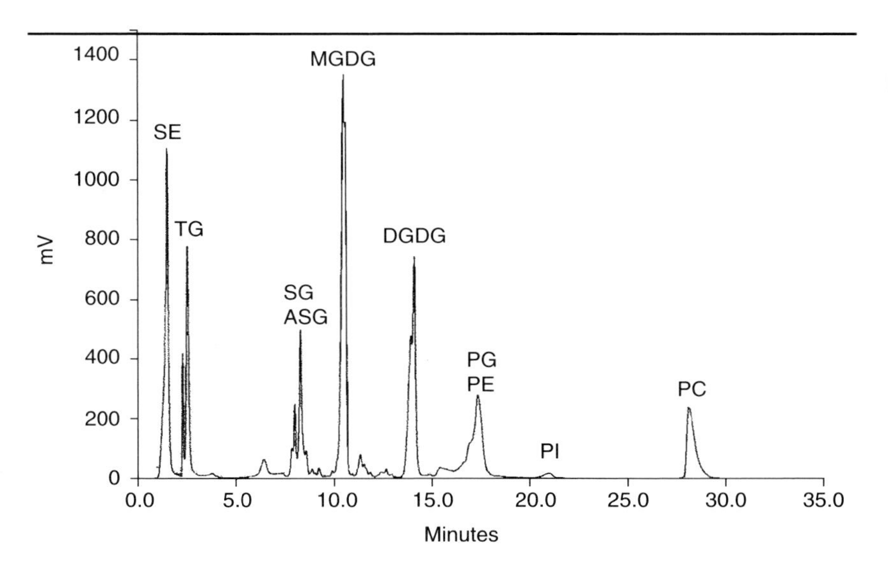

Separation of lipids from *Arabidopsis* leaf tissue by HPLC

Lipids were extracted from the leaf tissue of individual plants and run over the HPLC using a three gradient system and detected with an evaporative light-scattering device. Abbreviations not defined in the text: SE, sterol esters; TG, triacylglycerols; SG, steryl glycosides; ASG, acylsteryl glycosides; MGDG, monogalactosyldiacylglycerols; PG, phosphatidylglycerol; PI, phosphatidylinositol.

using mass spectrometry. The individual peaks probably contain a number of molecular species that will add to the puzzle. The isolation of specific mutants will allow us to understand the mechanisms that determine the complex lipid compositions of particular membranes and the hierarchy of complex lipids in the cell. They will also allow us to elucidate the fundamental roles of complex lipids in the metabolism and physiology of the whole plant.

The lesson learned from these ongoing studies is that collaboration with experts is essential to open up new areas of research.

A.R.S. wishes to thank BBSRC and Nickerson Biocem for their support in this research and the expertise of Toon Stuitje (Free University of Amsterdam) which was invaluable for learning how to handle E. coli mutants.

References

1. Ganguly, J. and Smellie, R.M.S., eds. (1972) Current Trends in the Biochemistry of Lipids (Biochemical Society, Symposium No. 35), Academic Press, London and New York
2. Stumph, P.K., ed. (1980) Lipids: Structure and Function, Vol. 4. The Biochemistry of Plants: A Comprehensive Treatise, Academic Press, New York
3. Farmer, E.E. and Ryan, C.A. (1990) Proc. Natl. Acad. Sci. U.S.A. **87**, 7713–7716
4. Spaink, H.P., Sheeley, D.M., van Brussel, A.A.N., Glushka, J., York, W.S., Tak, T., Geiger, O., Kennedy, E.P., Reinhold, V.N. and Lugtenberg, B.J.J. (1991) Nature (London) **354**, 125–130
5. Maurel, C., Reizer, J., Schroeder, J.I. and Chrispeels, M.J. (1993) EMBO J. **12**, 2241–2247
6. Wada, H. and Murata, N. (1989) Plant Cell Physiol. **30**, 971–978
7. de Mendoza, D. and Cronan, Jr., J.E. (1983) Trends Biochem. Sci. **8**, 49–52
8. Lundborg, T., Sandelius, A. S., Widell, S., Larsson, C., Liljenberg, C. and Kylin, A. (1982) in Biochemistry and Metabolism of Plant Lipids (Wintermans, J.F.G.M. and Kuiper, P.J.C., eds.), pp. 133–136, Elsevier Biomedical Press, Amsterdam
9. Rothman, J.E. and Wieland, F.T. (1996) Science **272**, 227–234
10. Cote, G.G. and Crain, R.C. (1993) Annu. Rev. Plant Physiol. Plant Mol. Biol. **44**, 333–356
11. McConn, M. and Browse, J. (1996) Plant Cell **8**, 403–416
12. Darnell, J., Lodish, H. and Baltimore, D., eds. (1990) Molecular Cell Biology, p. 501, Scientific American Books, Inc., New York
13. Vigh, L., Los, D.A., Horvatah, I. and Murata, N. (1993) Proc. Natl. Acad. Sci. U.S.A. **90**, 9090–9094
14. Knauf, V.C. (1987) Trends Biotechnol. **5**, 40–47
15. Somerville, C. and Browse, J. (1991) Science **252**, 80–87
16. Shorrosh, B.S., Dixon. R.A. and Ohlrogge, J.B. (1994) Proc. Natl. Acad. Sci. U.S.A. **91**, 4323–4327
17. Elborough, K.M., Swinhoe, R., Winz, R., Kroon, J.T.M., Farnsworth, L., Fawcett, T., Martinez-Rivas, J.M. and Slabas, A.R. (1994) Biochem. J. **301**, 599–605
18. Slabas, A.R., Fawcett, T., Griffiths, G. and Stobard, K. (1993) in Biosynthesis and Manipulation of Plant Products (Grierson, D., ed.), pp. 104–138, Blackie Academic and Professional, London
19. Pollard, M.R., Anderson, L., Fan, C., Hawkins, D.J. and Evans, H.M. (1991) Arch. Biochem. Biophys. **284**, 306–312
20. Voelker, T.A., Worrell, A.C., Anderson, L., Bleibaum, J., Fan, C., Hawkins, D.J., Radke, S.E. and Davies, H.M. (1992) Science **257**, 72–74
21. Arondel, V., Lemieux, B., Hwang, I., Gibson, S., Goodman, H.M. and Somerville, C. (1992) Science **258**, 1353–1355
22. Yadav, N.S., Wierzbicki, A., Aegerter, M., Caster, C.S., Perez-Grau, L., Kinney, A.J., Hitz, W.D., Booth, Jr., J.R., Schweiger, B., Stecca, K.L., Allen, S.M., Blackwell, M., Reiter, R.S., Carlson, T.J., Russell, S.H., Feldmann, K.A., Pierce, J. and Browse, J. (1993) Plant Physiol. **103**, 467–476
23. Stymne, S. and Appelqvist, L.-A. (1978) Eur. J. Biochem. **90**, 223
24. Rafferty, J.B., Simon, J.W., Stuitje, A.R., Slabas, A.R., Fawcett, T. and Rice, D.W. (1994) J. Mol. Biol. **237**, 240–242
25. Lindqvist, Y., Huang, W.J., Schneider, G. and Shanklin, J. (1996) EMBO J. **15**, 4081–4092
26. Baldock, C., Rafferty, J.B., Sedelnikova, S.E., Baker, P.J., Stuitje, A.R., Slabas, A.R. and Rice, D.W. (1996) Science **274**, 2107–2110

27. Newman, T., Debruijn, F.J., Green, P., Keegstra, K., Kende, H., McIntosh, L., Ohlrogge, J., Raikhel, N., Somerville, S., Thomashow, M., Retzel, E. and Somerville, C. (1994) Plant Physiol. **106**, 1241–1255
28. van de Loo, F.J., Turner, S. and Somerville, C. (1995) Plant Physiol. **108**, 1141–1150
29. Moore, T.S. (1982) Annu. Rev. Plant Physiol. **33**, 235–259
30. Paltauf, F., Kohlwein, S.D. and Henry, S.A. (1992) in Molecular Biology of the Yeast Saccharomyces (E.W. Jones, J.R. Pringle and J.R. Broach, eds.), pp. 4–97, Cold Spring Harbor Press, Cold Spring Harbor, NY
31. Nishida, I., Swinhoe, R., Slabas, S.R. and Murata, N. (1996) Plant Mol. Biol. **31**, 205–211
32. Dewey, R.E., Wilson, R.F., Novitzky, W.P. and Goode, J.H. (1994) Plant Cell **6**, 1495–1507
33. Coleman, J. (1990) J. Biol. Chem. **265**, 17215–17221
34. Harder, M.E., Ladenson, R.C., Schimmel, S.D. and Silbert, D.F. (1974) J. Biol. Chem. **249**, 7468–7475
35. Bell, R.M. (1974) J. Bacteriol. **117**, 1065–1076
36. Frisch, D.F., Tommey, A.M., Gegenbach, B.G. and Somers, D.A. (1991) Mol. Gen. Genet. **228**, 287–293
37. Snustad, D.P., Hunsperger, J.P., Chereskin, B.M. and Messing, J. (1988) Genetics **120**, 1111–1124
38. Brown, A.P., Coleman, J., Tommey, A.M., Watson, M.D. and Slabas, A.R. (1994) Plant Mol. Biol. **26**, 211–223
39. Verwoert, I., Brown, A., Slabas, A.R. and Stuitje, A.R. (1995) Plant Mol. Biol. **27**, 629–633
40. Hausmann, L., Schell, J. and Topfer, R. (1995) in Plant Lipid Metabolism (Kader, C. and Mazliak, P., eds.), pp. 534–536, Kluwer Academic, The Netherlands
41. Cao, Y., Oo, K. and Huang, A.H.C. (1990) Plant Physiol. **94**, 1199–1206
42. Brown, A.P., Brough, C.L., Kroon, J.T.M. and Slabas, A.R. (1995) Plant Mol. Biol. **29**, 267–278
43. Brough, C.L., Coventry, J.M., Christie, W.W., Kroon J.T.M., Brown, A.P., Barsby, T.L. and Slabas, A.R. (1996) Mol. Breed. **2**, 133–142
44. Ohlrogge, J. and Browse, J. (1995) Plant Cell **7**, 957–970
45. Dormann, P., Hoffmann-Benning, S., Balbo, I. and Benning, C. (1995) Plant Cell **7**, 1801–1810
46. Christie, W.W. and Urwin, R.A. (1995) High Resol. Chromatogr. **18**, 97–100

Cytochrome b_5 and polyunsaturated fatty acid biosynthesis

Mark A. Smith*, A. Keith Stobart*, Peter R. Shewry† and Johnathan A. Napier†‡
*School of Biological Sciences, University of Bristol, Bristol BS8 1UG, U.K. and †IACR-Long Ashton Research Station, Department of Agricultural Sciences, University of Bristol, Long Ashton, Bristol BS18 9AF, U.K.

Introduction

Although higher plants synthesize a diverse range of fatty acids, those with a chain length of 18 carbon atoms, containing one to three double bonds, predominate in crop plants grown for vegetable oil production. The number of these double bonds and their position in the acyl chain are major factors in determining the nature and economic importance of the oil. For this reason there is considerable interest in the manipulation of levels and types of unsaturated fatty acids in crop species such as sunflower and oil-seed rape to make oils more suitable for specific end purposes. For example, reduction of the polyunsaturated fatty acid (PUFA) content of a vegetable oil could result in a product with increased stability and consumer appeal, or one more suitable for the manufacture of margarine.

In the higher plant cell, fatty acids are synthesized *de novo* in the chloroplast (or plastid of non-green tissue). Unsaturated fatty acids are then formed by the sequential introduction of double bonds into a saturated precursor in a process catalysed by desaturase enzymes. These enzymes can only insert double bonds at single highly specific positions in the acyl chain and are selective for the substrate of desaturation. The first double bond is introduced while the precursor [(usually stearic acid (C18:0)] is attached to an acyl carrier protein in a reaction catalysed by a soluble, plastid-located enzyme. Further desaturation can then occur in one of two compartments within the cell, the plastid or the endoplasmic reticulum (ER) membrane. In most commercially important oil-seed crops the ER-located desaturase enzymes are responsible for the synthesis of PUFAs, usually linoleic acid (C18:2) and α-linolenic acid (C18:3), for storage in the form of triacylglycerol. These desaturases require fatty acids esterified to phosphatidylcholine as substrate and utilize an electron-transport chain consisting of cytochrome b_5 and NADH:cytochrome b_5 reductase (Figure 1).

Potential approaches to the manipulation of PUFA content in seed oils could therefore include targeting either specific desaturase enzymes or a

‡To whom correspondence should be addressed.

Figure 1

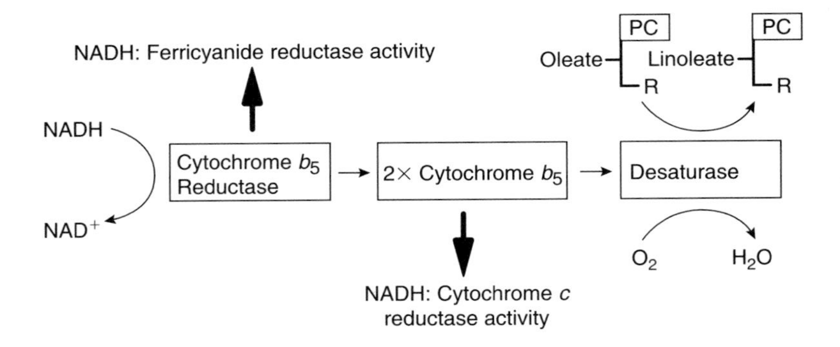

Scheme showing the electron transport components associated with the ER fatty acid desaturase enzymes of developing oil-seeds

Wide arrows indicate the assays that can be used to measure the activity of each component in vitro. PC = phosphatidylcholine, R = acyl group.

component of the electron transport system such as cytochrome b_5. To this end we have been using tobacco (*Nicotiana tabacum*), which has a seed oil containing over 70% linoleic acid (C18:2), as a model system to study the effects of altering cytochrome b_5 levels on PUFA biosynthesis. We have isolated a cDNA encoding cytochrome b_5 from a tobacco cDNA library, characterized the protein by expression in *E. coli* and produced transgenic plants containing either greatly increased or reduced levels of cytochrome b_5.

Cytochrome b_5: isolation of cDNA clones and biochemical characterization

Cytochrome b_5 is a haem protein composed of about 140 amino acids with a hydrophilic N-terminal domain to which a single haem group is bound, and a hydrophobic C-terminal region that acts as a membrane anchor. In animals it functions as an electron donor in a number of microsomal oxidation/reduction reactions, including the synthesis of cholesterol, the reduction of cytochrome *P*450s and the desaturation of acyl-CoA substrates (see [1] for references). In plants we have shown that cytochrome b_5 is the electron donor for the ER Δ^{12}-oleoyl-phosphatidylcholine desaturase (Δ^{12} desaturase) of developing safflower (*Carthamus tinctorius*) seeds [2] and is involved in the Δ^{12} hydroxylation of oleate to form ricinoleic acid in castor bean (*Ricinus communis*) [3]. The protein is also involved in the synthesis of sterols in plants.

To isolate a cDNA encoding tobacco microsomal cytochrome b_5, degenerate oligonucleotide primers were designed based on the amino acid sequences of cytochrome b_5 from animals and from cauliflower, the only sequence then available of a plant cytochrome b_5. These primers were used in PCR, with DNA isolated from a tobacco leaf cDNA library as a template, to isolate a DNA fragment encoding part of a cytochrome b_5 sequence. A cDNA encoding a full-length cytochrome b_5 was then obtained by screening the cDNA library using this fragment as a probe [4]. The deduced amino acid sequence of the protein showed 82% identity to the deduced amino acid sequence of cauliflower cytochrome b_5, and high similarity to animal microsomal cytochrome b_5 sequences (Figure 2). It also contained the conserved residues characteristic of the cytochrome b_5 family of proteins [5].

Further characterization of the tobacco cytochrome b_5 was conducted by expressing the cDNA in *E. coli*. This was achieved by transforming cells of *E. coli* strain BL21(DE3) with the expression vector pET-3d [6] containing either the complete coding region of the cDNA (F), or a truncated cDNA (T) encoding the soluble N-terminal domain of the protein (106 amino acids) only. Cells expressing either form of the protein were red in colour, exhibited an absorbance spectrum typical of reduced cytochrome b_5 and accumulated the recombinant protein to some 30% of total cellular protein (Figure 3). Cell fractionation suggested that the full-length recombinant cytochrome b_5 was present in inclusion bodies whereas the soluble form was located in the cytoplasm of the cells. The soluble form of the protein was purified for further characterization [1] and for use in the production of cytochrome b_5-specific polyclonal antibodies. The purified protein exhibited the characteristic absorbance spectra of plant microsomal cytochrome b_5 [2] (Figure 4) and was rapidly reduced by NADH in the presence of microsomal membranes prepared from developing sunflower seeds, indicating an ability to interact with the microsomal electron-transport systems. CD spectroscopy indicated that the protein contained around 35% α-helix.

Figure 2

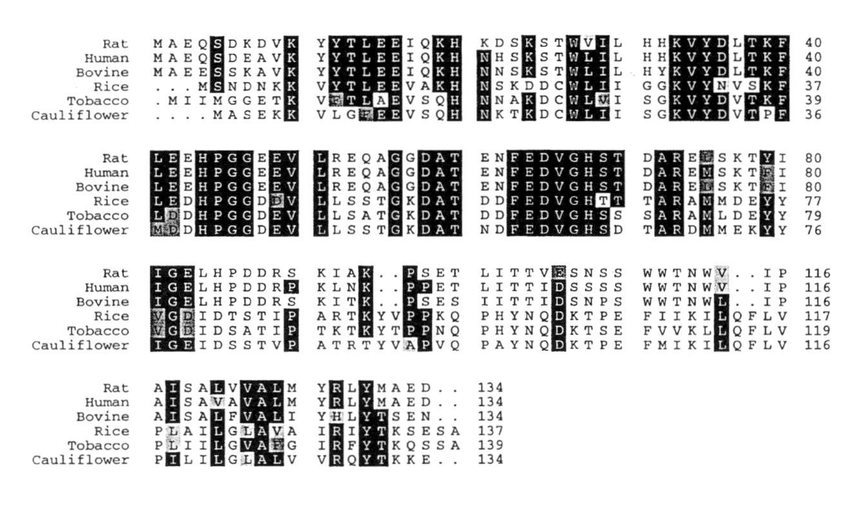

Comparison of the deduced amino acid sequences of mammalian and plant cytochrome b_5

Figure 3

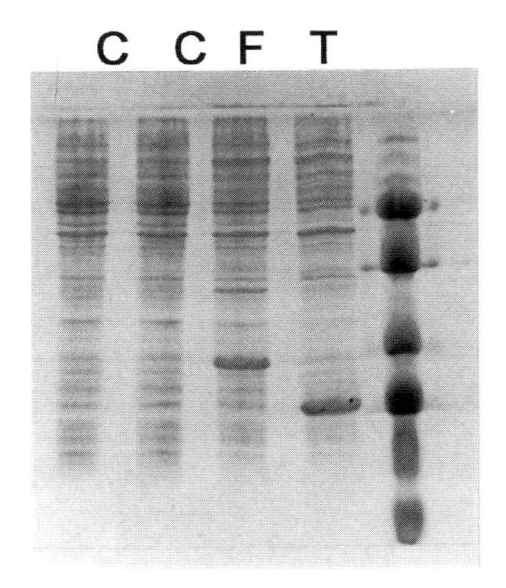

Coomassie-Blue-stained SDS/PAGE separation of total proteins from untransformed *E. coli* cells (C) and cells expressing the F and T forms of tobacco cytochrome b_5

Reproduced from [1] with permission from the Biochemical Society.

Figure 4

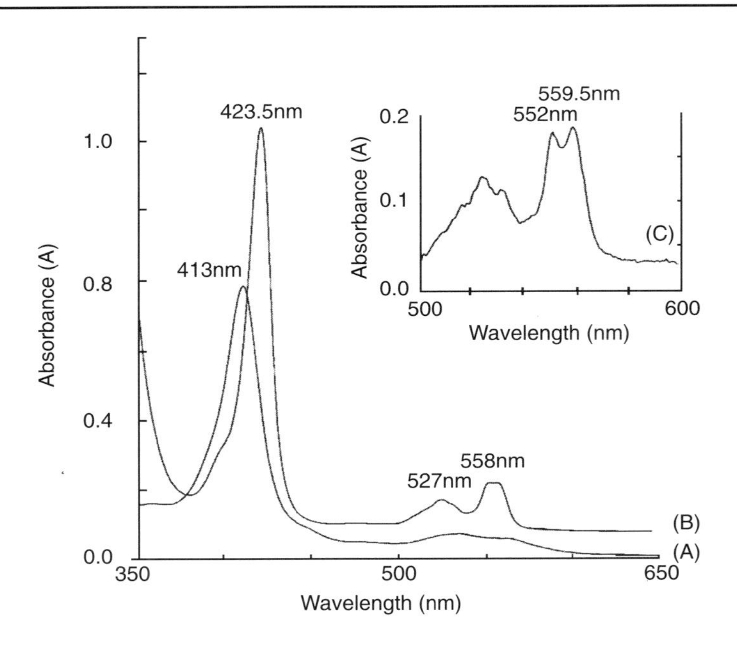

Absorbance spectra of an aqueous solution of the purified T form of cytochrome b_5

(A) Spectrum of air-oxidised cytochrome. (B) Spectrum of sodium dithionite-reduced cytochrome. (C) Low temperature (77 K) spectrum of sodium dithionite-reduced cytochrome, showing characteristic splitting of the α and β absorbance bands. Reproduced from [1] with permission from the Biochemical Society.

Manipulation of cytochrome b_5 levels in tobacco

Agrobacterium-mediated leaf disc transformation was used to introduce the following cDNAs into tobacco: (1) a cDNA encoding the full-length (F) tobacco cytochrome b_5; (2) a cDNA encoding the truncated (T) form of cytochrome b_5 lacking the C-terminal membrane-anchoring domain; and (3) a cDNA encoding the full-length tobacco cytochrome b_5 in reverse orientation (antisense).

Constructs were cloned into the binary vector pBin19 downstream of the CaMV-35S promoter and an Ω-translational enhancer [7].

Expression of the transgene in the successfully transformed tobacco plants was analysed by Northern blotting of total RNA extracted from leaf tissue (Figure 5). Plants containing the F transgene contained greatly increased amounts of cytochrome b_5 message. Those containing the T transgene showed similar levels of endogenous b_5 message to the control plants and in addition contained high levels of message of the size predicted for the T transcript. Levels of expression varied considerably between transformed lines (see Figure 5). Further analysis demonstrated that the transgenes were expressed in all tissues examined (leaf, seed, root and stem). The levels of cytochrome b_5 protein in the transformed plants were then assessed by Western blotting (Figure 6). Plants expressing the F form of the protein contained greatly increased levels of cytochrome b_5 while those expressing the T form contained the endogenous cytochrome b_5 and also accumulated the truncated form, although at a lower level than expected from the amount of message observed. Only one of the plants transformed with the cytochrome b_5 antisense construct showed a reduction in the level of cytochrome b_5 (Figure 6) and this was observed in leaf and root tissue but not in developing seeds.

To assess the effects of altering cytochrome b_5 levels on the electron transport activities of the plants, *in vitro* assays were conducted on microsomal membranes prepared from the roots of wild-type plants and selected transgenic plants that expressed the highest levels of the F and T forms of cytochrome b_5. The NADH:ferricyanide reductase activities of these plants were essentially the same,

Figure 5

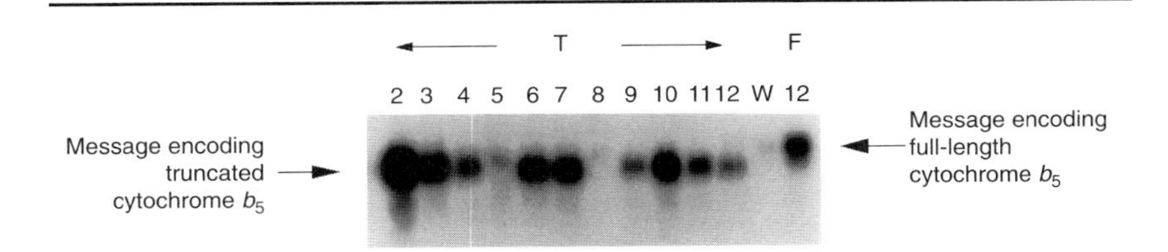

Northern blot of total RNA from leaves of a number of transformed plants expressing the truncated form of cytochrome b_5 (T), the full-length form of cytochrome b_5 (F) and the wild-type tobacco (W)

The blot was probed with radiolabelled cDNA encoding tobacco cytochrome b_5. Numbers refer to different plant lines.

Figure 6

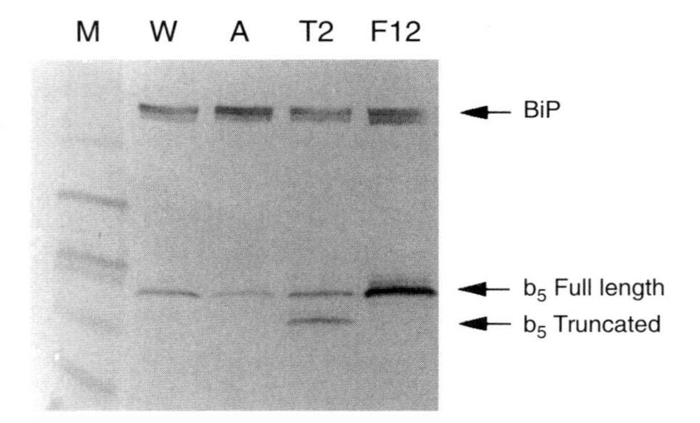

Western blot of total leaf proteins separated by SDS/PAGE from wild-type tobacco (W), plants expressing the (F) and (T) forms of cytochrome b_5, and a plant containing the cytochrome b_5 antisense construct (A)

The blot was probed with a polyclonal antibody raised against the TC1 form of tobacco cytochrome b_5 and an antibody against the ER binding protein BiP, which acts as a loading control. Numbers refer to different plant lines, M = molecular mass marker proteins.

with an average of 175 nmol·min^{-1} of ferricyanide reduced per mg of protein (Figure 7). This assay measures the activity of the NADH:cytochrome b_5 reductase component of the ER electron transport system (Figure 1) and indicated that the activity of the enzyme was not altered in the transgenic plants. Assays of NADH:cytochrome *c* reductase activity (a measure of electron flow through cytochrome b_5, see Figure 1) showed that plants expressing the T form of the protein had similar activities to control plants at 40–50 nmol·min^{-1} of cytochrome *c* reduced per mg of protein (Figure 7) whereas those expressing the F form had significantly increased activity. This observation indicates that the full-length

Figure 7

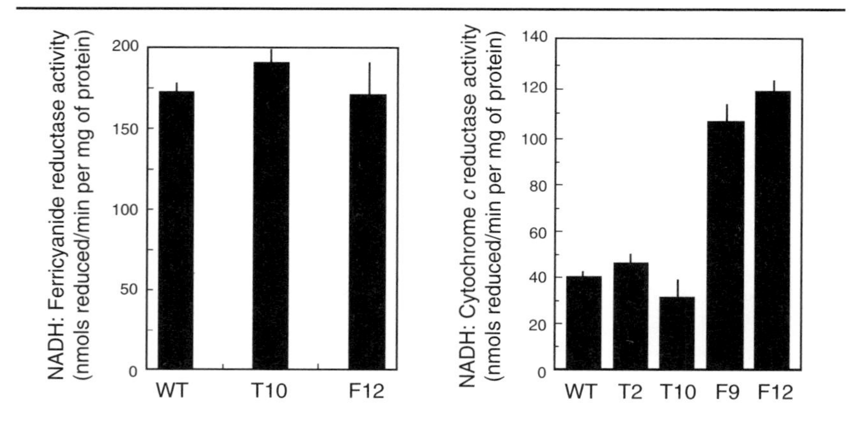

Reductase activities of microsomal membranes prepared from the roots of wild-type tobacco (WT) and plants expressing the (F) and (T) forms of cytochrome b_5

Numbers refer to different plant lines. For the error bars, n = 3.

cytochrome b_5 produced by expression of the transgene is able to function in electron transport in the membranes.

To determine the fatty acid profiles of the transformed plants, total lipids were extracted from leaves and mature seeds of individual plants, and fatty acid methyl esters were prepared and separated by gas chromatography (Figure 8). Although some variation was observed between the relative amounts of fatty acids in the plants, no significant differences were observed in the levels of any fatty acid in either the leaves or seeds of the transformants.

Discussion

Characterization of the cytochrome b_5 encoded by the tobacco cDNA described above indicates that it encodes a functional protein that is highly similar to the ER-located cytochrome b_5 proteins of animals and fungi. Expression of the cDNA in tobacco under the control of a constitutive promoter (CaMV-35S) resulted in plants that contained significantly higher levels of cytochrome b_5 than

Figure 8

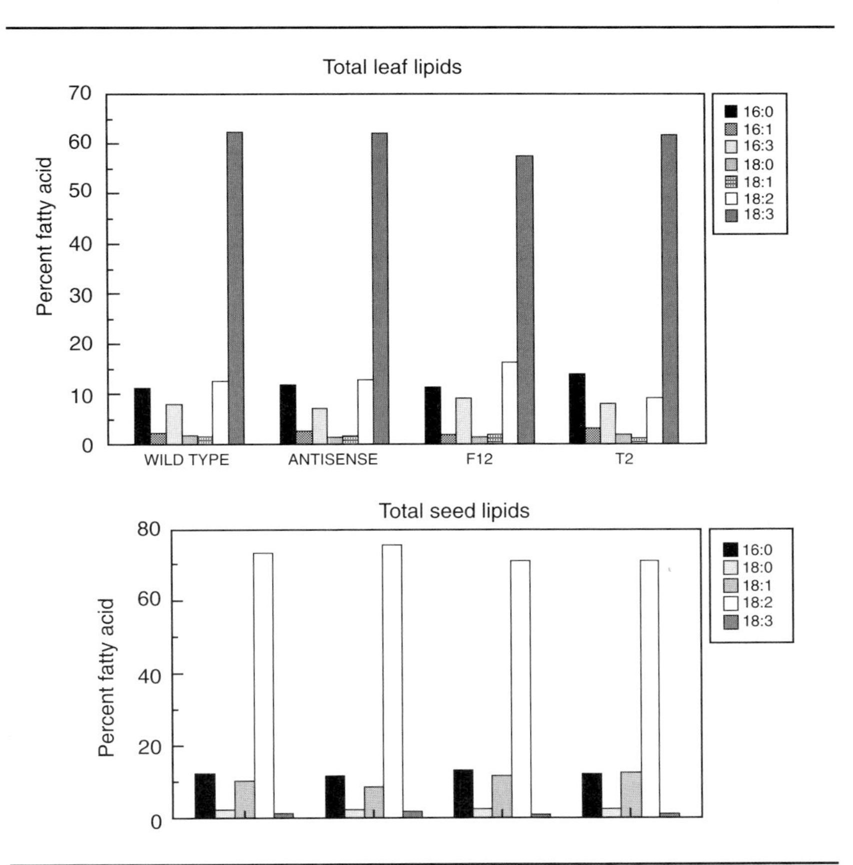

Fatty acid profiles of leaves and seeds of wild-type tobacco and plants with altered levels of cytochrome b_5

Plants F12 and T2 express the full-length form of cytochrome b_5 and the truncated form, respectively.

wild-type plants. Western blot analysis and assays of microsomal reductase activity demonstrated that the recombinant protein was correctly targeted to the ER and was able to function in electron transport. The plants, however, appeared phenotypically identical to wild-type plants and showed no difference in fatty acid content in any tissue analysed. These results would suggest that the amounts of cytochrome b_5 and electron transport to the ER desaturases do not limit PUFA synthesis in tobacco. The expression of a truncated form of cytochrome b_5 in tobacco resulted in similar findings. This form of the protein is not able to insert into the ER membrane although *in vitro* assays indicate that it is able to interact with microsomal cytochrome b_5 reductase and may divert reducing potential from the endogenous cytochrome b_5.

Expression of the cytochrome b_5 cDNA in an antisense orientation in tobacco resulted in significant reduction in the cytochrome b_5 levels in leaf tissue. The plants appeared phenotypically normal and showed no alteration in fatty acid profiles. This finding is perhaps not unexpected as in leaves the major source of PUFAs is thought to be the plastid where ferredoxin, not cytochrome b_5, acts as the electron donor to the desaturase enzymes. However, no significant reduction in the level of cytochrome b_5 was observed in the developing seed of the plant. Since our previous studies have shown that the CaMV-35S promoter is active in developing seed (using the same promoter construct we have expressed the T form of cytochrome b_5 in tobacco seeds) this suggests that the cytochrome b_5 expressed in the seed is different to that expressed in leaves. In fact we have recently isolated a cDNA encoding cytochrome b_5 that appears to be specifically expressed in developing seeds of tobacco [8], and we are now using this to continue our studies into the effect of lowering cytochrome b_5 levels on PUFA synthesis. It is therefore necessary to await the results of these studies before conclusions can be drawn on the feasibility of regulating fatty acid desaturation in oil-seeds by manipulating the expression of cytochrome b_5.

References

1. Smith, M.A., Napier, J.A., Stymne, S., Tatham, A.S., Shewry, P.R. and Stobart, A.K. (1994) Expression of a biologically active plant cytochrome b_5 in *Escherichia coli*. Biochem. J. **303**, 73–79
2. Smith, M.A., Cross, A.R., Jones, O.T.G., Griffiths, W.T., Stymne, S. and Stobart, A.K. (1990) Electron-transport components of the 1-acyl-2-oleoyl-*sn*-glycero-3-phosphocholine D12-desaturase from developing safflower (*Carthamus tinctorius* L.) cotyledons. Biochem. J. **272**, 23–29
3. Smith, M.A., Jonsson, L., Stymne, S. and Stobart, A.K. (1992) Evidence for cytochrome b_5 as an electron donor in ricinoleic acid biosynthesis in microsomal preparations from developing castor bean (*Ricinus communis* L.). Biochem. J. **287**, 141–144
4. Smith, M.A., Stobart, A.K., Shewry, P.R. and Napier, J.A. (1994) Tobacco cytochrome b_5: cDNA isolation, expression analysis and *in vitro* protein targeting. Plant Mol. Biol. **25**, 527–537
5. Mathews, F.S. (1985) The structure, function and evolution of cytochromes. Prog. Biophys. Mol. Biol. **45**, 1–56
6. Studier, F.W., Rosenberg, A.H., Dunn, J.J. and Dubendorff, J.W. (1990) Use of T7 polymerase to direct expression of cloned genes. Methods Enzymol. **185**, 60–89
7. Gallie, D.R., Lucas, W.J. and Walbot, V. (1989) Visualising mRMA expression in plant protoplasts: Factors influencing efficient mRNA uptake and translation. Plant Cell. **1**, 301–311
8. Napier, J.A., Smith, M.A., Stobart, A.K. and Shewry, P.R. (1995) Isolation of a cDNA encoding a cytochrome b_5 specifically expressed in developing tobacco seeds. Planta **197**, 200–202

Bioengineering terpenoid production: current problems

D. McCaskill and R. Croteau*
Institute of Biological Chemistry, Washington State University, Pullman, WA 99164-6340, U.S.A.

Introduction

Terpenoids are distributed throughout nature and have been exploited by humans throughout history for a myriad of uses. The commercial uses of terpenoids include the industrial processing of low-value, high-volume products such as the turpentine isolated from conifer oleoresin, as well as the isolation of very high-value, low-volume products such as terpenoid-derived pharmaceuticals. One of the most important commercial uses of terpenoids is the utilization of turpentine as the starting material in the industrial production of synthetic high-value end products [1]. The most common constituents of the essential oils, which have been valued throughout history for their aromatic and culinary uses, are terpenoids (see, for example, [2]). Many pharmaceuticals that are either in current use or which show clinical promise include terpenoids such as taxol [3,4], artemisinin [5], forskolin [6] and ginkgolides [7], which are used in the treatment of cancer, malaria, glaucoma and cardiovascular disorders, respectively. The vast majority of the over 25 000 known terpenoids found in nature [8] are produced by plants. In addition to the role that terpenoids such as sterols and carotenoids play in cell growth and development, plants rely on many terpenoids for defence against herbivores and pathogens, allelopathic interactions and attraction of pollinators (reviewed in [9,10]). Thus, an intimate understanding of the enzymology and regulation of terpenoid biosynthesis in plants has broad implications for the improvement of: the industrial use of terpenoid feedstocks derived from plants; the discovery of novel pharmaceuticals; the improved production of existing pharmaceuticals; the development of novel flavourings and fragrances; and the development of improved disease and pest resistance of commercially important crop plants.

The following review will focus on two areas of terpenoid biosynthesis that are currently undergoing intense investigation. The first section will discuss the current state of knowledge of isopentenyl diphosphate (IPP) biosynthesis. IPP serves as the fundamental biosynthetic building block for the biosynthesis of all terpenoids. However, despite the importance and ubiquitous distribution of terpenoids, many fundamental questions remain concerning the biosynthesis of IPP in plants. The second section will focus on recent developments in the isolation, characterization and cDNA cloning of the enzymes responsible for the synthesis of selected terpenoids.

*To whom correspondence should be addressed.

The pathways and subcellular compartmentation of IPP biosynthesis

IPP represents a crucial branch-point metabolite in the biosynthesis of terpenoids (Figure 1). Plant cells must balance and regulate the partitioning of IPP into different families of terpenoids, including sterols, carotenoids, hormones and secondary products such as monoterpenes, sesquiterpenes and diterpenes, often within the same cell. The role of the mevalonic acid (MVA) pathway in producing IPP for terpenoid biosynthesis is widely accepted and will not be reviewed in detail here. The enzymology [11,12] and subcellular organization of the MVA pathway [12] have been recently reviewed. A major focus of research on the regulation of terpenoid biosynthesis has been directed at the role that 3-hydroxy-3-methylglutaryl-coenzyme A reductase (HMGR) plays, and this regulatory role will be briefly discussed. Recent studies demonstrating a novel, MVA-independent pathway for terpenoid biosynthesis will also be discussed in detail.

HMGR has been extensively studied because of its pivotal role in regulating cholesterol biosynthesis in animals [13]. Numerous studies have demonstrated that HMGR is also heavily regulated in plants (reviewed in [14,15]). In all plants examined to date, HMGR is encoded by small gene families. The number of HMGR genes present varies from only two in *Arabidopsis* to over 12 in the case of potato [14]. Research on the differential expression of these different isoforms of HMGR in different tissues and in response to developmental changes

Figure 1

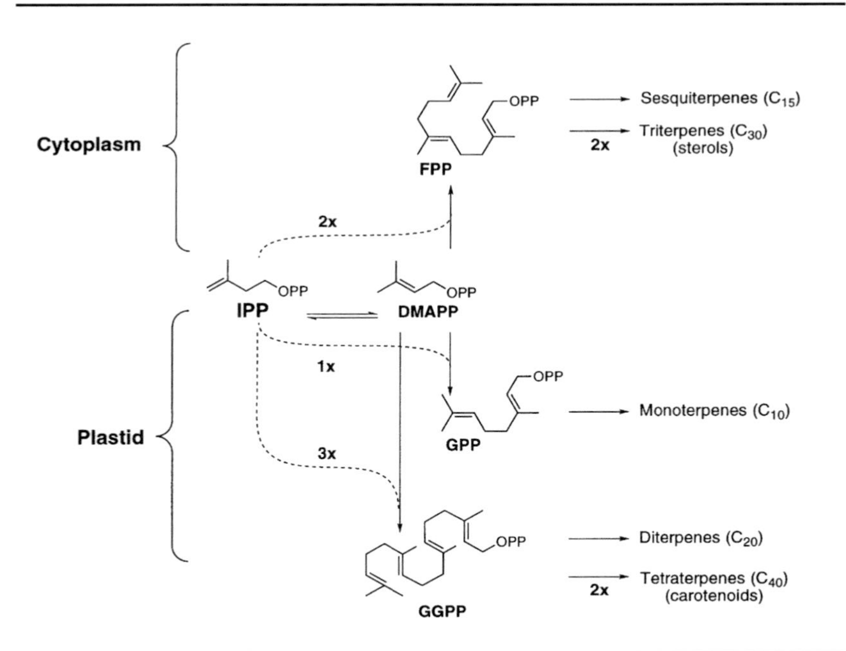

Biosynthesis of the major classes of isoprenoids in plants from IPP

The generally accepted subcellular compartmentation of the prenyltransferases GPP synthase, FPP synthase and GGPP synthase is shown. The diphosphate moiety is abbreviated as –OPP. DMAPP, dimethylalkyl diphosphate; FPP, farnesyl diphosphate; GGPP geranylgeranyl diphosphate; GPP, geranyl diphosphate.

and environmental challenges has focused on the role of these isoforms in regulating the biosynthesis of different families of terpenoids. In particular, the role of HMGR in regulating the biosynthesis of terpenoid-derived phytoalexins has received considerable attention (reviewed in [15]). Despite this attention, the subcellular compartmentation of HMGR, and the role it may play in regulating terpenoid biosynthesis, is still controversial.

All plant HMGR genes examined to date encode two highly conserved hydrophobic membrane-spanning regions near the N-terminus of the protein, and a highly variable N-terminal sequence of varying length. The conserved hydrophobic N-terminal sequences have recently been shown to serve as an internal signal sequence for retention of HMGR in the endoplasmic reticulum (ER) [16]. The topology of HMGR has been demonstrated to consist of these two membrane-spanning regions, which anchor HMGR to the ER, with the N-terminus and the C-terminal domain (containing the catalytic site) facing the cytoplasm [16,17]. A short sequence between the hydrophobic membrane-spanning regions is exposed to the lumen of the ER. An interesting observation is that the HMGR isoforms that are associated with the production of elicited defensive compounds, or other secondary products, contain an N-linked glycosylation site in this short lumenal sequence, whereas HMGR that is constitutively expressed, or which is associated with sterol formation, lacks this glycosylation site [17]. Thus, glycosylation of specific forms of HMGR may play an important role in the localization or regulation of HMGR activity.

The role of the MVA pathway (Figure 2) as the exclusive source of IPP in plants has been called into question recently by the demonstration of novel labelling patterns in terpenoids derived from ^{13}C-labelled substrates in a variety of bacteria, a green alga and in plant tissue cultures: these labelling patterns are inconsistent with the operation of the MVA pathway. The presence of a novel non-mevalonate pathway was originally demonstrated in a variety of eubacteria, including *Escherichia coli* [18], by assessing the labelling pattern of hopanoids and ubiquinones derived from ^{13}C-labelled glucose, acetate and pyruvate. The

Figure 2

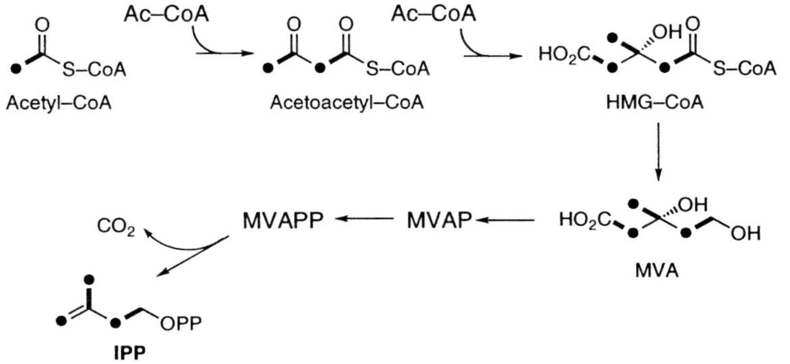

Biosynthesis of IPP by the MVA pathway

The labelling pattern of IPP from acetyl-CoA is indicated. MVAP, mevalonate 5-phosphate; MVAPP, mevalonate 5-diphosphate; HMG, 3-hydroxy-3-methylglutaric acid.

labelling pattern observed in these terpenoids (expressed by conceptually breaking the terpenoids down into their constituent C_5 units originally derived from IPP) is inconsistent with the operation of the MVA pathway in these bacteria. The authors concluded that the branched C_5 skeleton of IPP was derived from the combination of a C_3 substrate derived from glycolysis and an activated C_2 intermediate. Subsequent work with a series of *E. coli* mutants, each of which is defective in a single enzyme of triose phosphate metabolism [19], has led to the proposal that glyceraldehyde 3-phosphate (GAP) and an activated C_2 intermediate derived from pyruvic acid are the two substrates that are utilized to generate IPP via this novel pathway (Figure 3). The presence of this 'GAP/pyruvate' pathway has recently been demonstrated in the green alga *Scenedesmus obliquus* [20]. In this case, the incorporation of [U-^{13}C]- and [4,5-^{13}C]glucose into the terpenoids examined resulted in characteristic ^{13}C–^{13}C coupling constants, indicating that an intact C_2 unit derived from glucose is incorporated into the C-3 and C-5 carbons of IPP, and that an intact C_3 unit derived from glucose is incorporated into the C-1, C-2 and C-4 carbons of IPP. In this study, the carotenoids, sterols and the prenyl side chains both of chlorophylls and plastoquinone were all labelled by this novel pathway, indicating that at least some eukaryotes may rely entirely on this non-mevalonate pathway for terpenoid biosynthesis.

The operation of the GAP/pyruvate pathway has also been demonstrated in plants, although the distribution and importance of the pathway for terpenoid biosynthesis is currently not clear. Using germinating embryos of *Ginkgo biloba*, the labelling patterns of the diterpene ginkgolides were shown to be consistent with the GAP/pyruvate pathway, whereas sterols were labelled in accordance with the mevalonate pathway [21,22]. More recently, the GAP/pyruvate pathway has been implicated in the biosynthesis of the diterpenoid taxane skeleton of taxayunnanine C in tissue cultures of *Taxus chinensis* [23]. The ^{13}C–^{13}C couplings of taxayunnanine C derived from [U-^{13}C]glucose convincingly demonstrate that plant cells are capable of synthesizing

Figure 3

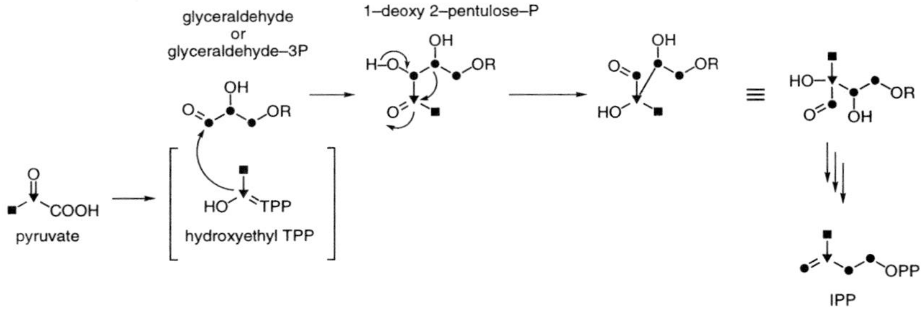

Postulated biosynthesis of IPP by the GAP/pyruvate pathway

The fate of each carbon derived from pyruvate and either glyceraldehyde or GAP is indicated and is based on labelling experiments described in the text. Conversion of the initial branched isopentanoid intermediate into IPP should involve three reductions, three dehydrations and one or more phosphorylations. R = H or PO_4^{2-}.

plastid-derived diterpenes via this novel GAP/pyruvate pathway. These studies would seem to suggest that the classical mevalonate pathway is responsible for the cytoplasmic synthesis of triterpene sterols (and presumably sesquiterpenes) derived from farnesyl diphosphate (FPP) whereas the GAP/pyruvate pathway may be responsible for the synthesis of plastid-derived terpenoids such as diterpenes. Although the GAP/pyruvate pathway has only been implicated in the synthesis of diterpenes in higher plants at this time, other plastid-derived terpenoids such as monoterpenes and carotenoids may also rely on IPP produced by this novel pathway. However, it cannot be assumed *a priori* that the GAP/pyruvate pathway is universally distributed in plants or that it is exclusively responsible for the synthesis of plastid-derived terpenoids. In contrast to the studies with *Scenedesmus* [20], *Gingko* [21,22] and *Taxus* [23], recent work with cultured cells of the photosynthetic liverwort *Heteroscyphus planus* has demonstrated the incorporation of [^{13}C]acetate into clerodane-type diterpenoic acids (e.g. heteroscyphic acid A), as well as cadinane-type sesquiterpenes via the classical MVA pathway [24]. However, the diagnostic experiments of assessing the coupling constants from the incorporation of the intact C_2 and C_3 units from [U-^{13}C]glucose were not carried out because this work was not specifically directed at assessing the presence or absence of the GAP/pyruvate pathway.

Several lines of evidence clearly demonstrate that plastids are capable of supporting terpenoid biosynthesis with plastid-derived IPP. Numerous studies have demonstrated a differential sensitivity of sterol, carotenoid and ubiquinone biosynthesis towards the HMGR inhibitors mevinolin and compactin (reviewed in [25–27]). Although mevinolin very effectively inhibits sterol formation *in vivo*, it is not an effective inhibitor of either carotenoid or ubiquinone formation, indicating the presence of at least two subcellular compartments for terpenoid biosynthesis, one of which is either inaccessible or insensitive to mevinolin. The recent demonstration of the GAP/pyruvate pathway in green alga [20] and higher plants [21–23] has been used to argue that this mevinolin-insensitive formation of plastidic terpenoids is due to this novel pathway, and not owing to the presence of a plastid-localized MVA pathway. However, the presence of a plastidic MVA pathway cannot yet be discounted. Numerous reports have described a plastidic form of HMGR activity that exhibits regulatory and kinetic behaviour distinct from the microsomal form of the enzyme (e.g. [28] and reviewed in [29]). Developmental changes in the preferred substrate for chloroplast terpenoid biosynthesis have also been observed in developing barley leaves [30]. In young tissue that contains immature chloroplasts, a high level of incorporation into chloroplast terpenoids is observed from $NaH^{14}CO_3$, whereas exogenous [^{14}C]MVA yielded little incorporation. However, during chloroplast maturation, plastidic IPP biosynthesis from $NaH^{14}CO_3$ drops, with a concomitant increase in incorporation from exogenous MVA. These results indicate that a developmental shift in the source of IPP utilized for chloroplast terpenoid biosynthesis occurs during chloroplast maturation. Thus, immature chloroplasts appear to be autonomous for IPP biosynthesis from CO_2, whereas mature chloroplasts rely on the import of cytoplasmic IPP. The presence of a stromal pathway for conversion of 3-phosphoglycerate into acetyl-CoA in purified barley chloroplasts indicates that a flow of carbon from CO_2 to acetyl-CoA utilized for terpenoid biosynthesis is possible [31]. Isolated, immature chloroplasts from spinach exhibit high rates of terpenoid biosynthesis from both [^{14}C]acetate and [^{14}C]pyruvate after hypotonic treatment, although rates of incorporation from acetate were approximately 5-fold higher than from pyruvate [32]. The high incorporation from [^{14}C]acetate argues against the involvement of the GAP/pyruvate pathway in these immature chloroplasts. Hypotonic treatment of immature barley chloroplasts also reveals latent activities of mevalonate kinase and mevalonate-phosphate kinase [33]. Collectively, these studies support the presence of a plastidic MVA pathway, albeit one that may be active only during a short period of plastid development.

Further complicating the field of IPP biosynthesis are a series of studies carried out with tissue cultures of mulberry (*Morus alba*) that accumulate large amounts of the prenylated-chalcone chalcomoracin. The incorporation of [U-^{13}C]glucose into the hemiterpene moieties of chalcomoracin [34,35], and of [^{13}C]acetate into sterols [36], are consistent with the classical mevalonate pathway. However, exogenous [2-^{13}C]acetate is incorporated into the hemiterpene moieties only indirectly, after conversion into [1,2-^{13}C]acetate by passage through the tricarboxylic acid cycle. This endogenously generated [1,2-^{13}C]acetate is incorporated into only the first acetate unit of the hemiterpene moieties. The incorporation of [1,3-^{13}C]- and [2-^{13}C]glycerol clarified this asymmetric labelling of the IPP units [34,35]. The model established by these studies indicates that, although the hemiterpene moieties are formed by the MVA pathway, the first acetate unit incorporated into IPP is derived directly from glycolysis and the second and third acetate units are derived from the pentose phosphate cycle. Additionally, the incorporation of [2-^{13}C]acetate into the hemiterpene moieties is unaffected by elevated levels of the HMGR inhibitor compactin under conditions in which incorporation of [^{13}C]acetate into sterols is eliminated [37]. This strongly suggests the presence of the MVA pathway in at least two distinct compartments, one of which is inaccessible to compactin. However, it is difficult to visualize how two distinct pools of acetyl-CoA, one derived from glycolysis and the other from the pentose phosphate cycle, can both be utilized by the MVA pathway within a single subcellular compartment in a way that leads to an asymmetric labelling of IPP. The same conceptual problem has been encountered in rationalizing a lack of competition observed for acetate (or acetyl-CoA) between plastidic terpenoid and acyl lipid biosynthesis (reviewed in [38,39]). Recent work, however, has demonstrated that acyl lipid biosynthesis in permeabilized chloroplasts tightly channels acetate into acyl lipids, and that acetyl-CoA neither competes with acetate nor is incorporated itself into acyl lipids [40]. This work indicates that the operationally soluble enzymes of acyl lipid biosynthesis are tightly organized into a multienzyme complex, providing a basis for the presence of distinct and separate pools of acetyl-CoA that are utilized for acyl lipid and terpenoid biosynthesis within the plastid.

Although no enzymology of the GAP/pyruvate pathway has been carried out yet, several predictions can be made based on the labelling patterns observed. The activated C_2 unit is probably generated by the decarboxylation of pyruvic acid, resulting in the formation of an enzyme-bound hydroxyethyl-thiamine diphosphate adduct similar to that formed by pyruvate dehyrogenase (Figure 2). The resulting activated hydroxyethyl adduct can then condense with the aldehyde of either GAP, generating a 1-deoxypentulose 5-phosphate, or with glyceraldehyde to generate a 1-deoxypentulose. It is worth noting that 1-deoxylulose is a precursor of both pyridoxol [41,42] and thiamine diphosphate [43] and that 1-deoxyxylulose is formed from pyruvate and glyceraldehyde by cell-free extracts of *E. coli* and *Bacillus subtilis* [44]. A carbon–carbon bond migration subsequently converts the linear deoxypentulose into the branched C_5 isopentanoid skeleton of IPP. A series of three dehydrations, three reductions and one or two phosphorylations could then generate IPP. However, there is no reason to believe *a priori* that the final product of this series of reactions will be IPP. Depending on the order of the dehydrations and reductions, the GAP/pyruvate pathway could as readily produce dimethylallyl diphosphate (DMAPP) as the final product, which would then equilibrate with IPP through the action of IPP isomerase.

The labelling patterns of terpenoids derived from stable isotopic precursors have convincingly demonstrated the presence of the GAP/pyruvate pathway for IPP formation in higher plants [21–23], as well as metabolic subtleties in the formation of IPP from acetate by the classical MVA pathway [34,35,37]. These studies illustrate the power of stable isotope techniques for elucidating metabolic pathways *in vivo*; however, they now need to be complemented with enzymological and molecular biological approaches. As discussed earlier, all HMGR isoforms examined to this point contain two highly conserved hydrophobic membrane-spanning regions that serve as internal signal sequences for retention in the ER [16]. However, the possibility cannot be excluded that certain isoforms of HMGR may be targeted to plastids or mitochondria after insertion into the ER, or that other residues present at the N-terminus of certain isoforms may play a role in targeting of HMGR to different parts of the endomembrane system [16]. The only other gene encoding an enzyme of the mevalonate pathway in plants before IPP that has been defined is mevalonate kinase from *Arabidopsis* [45]. The deduced amino acid sequence of this enzyme contains no putative plastid targeting sequence, and Southern blot analysis suggested the presence of only one locus for this gene, arguing against the presence of a plastidic MVA pathway in *Arabidopsis*. Other than acetyl-CoA acetyltransferase (producing acetoactetyl-CoA), HMG-CoA synthase [46,47] and HMGR [29,48–51], no detailed characterization of the enzymes of the MVA pathway in plants has been carried out.

It is unlikely that broad generalizations regarding the source of IPP utilized for biosynthesis of a particular class of terpenoid can be made. Regardless of the biosynthetic pathway utilized for IPP biosynthesis, several studies suggest that there may not be a strict subcellular segregation of IPP partitioning into different families of terpenoids. A developmental switch from autonomous formation of IPP by immature chloroplasts to a reliance on imported IPP for plastidic terpenoid formation [30,32] has already been discussed. Studies with semipermeable isolated secretory cells from glandular trichomes of peppermint demonstrated that both plastidial monoterpene and cytoplasmic sesquiterpene biosynthesis rely exclusively on plastid-derived IPP [52]. Based on this work, the low level of sesquiterpenes relative to monoterpenes found in the essential oils of many plants was suggested to be due to the preferential utilization of plastid-derived IPP by the plastidic monoterpene biosynthetic enzymes at the expense of cytoplasmic sesquiterpene biosynthesis. Experiments with these semipermeable secretory cells have also demonstrated high incorporation rates of exogenous geranyl diphosphate (GPP) into monoterpenes [52,53], indicating efficient uptake of GPP into the leucoplasts of these cells. Non-equivalent labelling by exogenous [^{13}C]MVA of the C_5 units of diterpenes has recently been observed with ginkgolides [22] and heteroscyphic acid A [24]. The labelling patterns of these diterpenes demonstrated incorporation of exogenous [^{13}C]MVA into the first three C_5 units of geranylgeranyl diphosphate (GGPP), with no incorporation into the terminal C_5 unit. This was interpreted as resulting from uptake of cytoplasmic-derived ^{13}C-labelled FPP into the plastids, elongation to GGPP by the addition of an unlabelled, plastid-derived IPP unit, and subsequent conversion into the labelled diterpenes [22,24]. The physiological relevance of these observations is unclear. Developmental changes in plastid envelope permeability towards terpenoid precursors such as acetate, mevalonate and IPP have been observed [30,54,55]. Collectively, these studies indicate that transport of substrates between the cytoplasm and plastid may play a role in the partitioning of IPP between different families of terpenoids and in the developmental regulation of terpenoid biosynthesis. A transport system for IPP in isolated plastids from *Vitis vinifera* L. cell suspension cultures has also been reported [56], although the apparent K_m for IPP (approx. 0.5 mM) is so high that the physiological role of such transport is unclear.

The picture that emerges from these recent studies is that plants possess parallel, independent pathways for the biosynthesis of IPP in different subcellular locales. The discrete labelling patterns observed for different classes of terpenoids clearly demonstrate that the pools of IPP derived from the classical MVA pathway and the GAP/pyruvate pathway have different metabolic fates. For several reasons, however, the picture is not yet complete. The presence of both the classical MVA pathway and the GAP/pyruvate pathway for IPP biosynthesis within plastids cannot yet be ruled out. If both pathways do exist within plastids, the developmental interplay between these pathways must be established. There is a pressing need to assess the taxonomic, tissue and developmental distribution of the GAP/pyruvate pathway. Developmental changes in the ability of developing chloroplasts, as well as non-photosynthetic plastids, to produce and transport the IPP necessary to support terpenoid biosynthesis need to be systematically examined. The labelling patterns of other families of terpenoids, including monoterpenes, carotenoids and ubiquinones, also need to be assessed. Elucidation of the enzymology, and ultimately the regulation, of the GAP/pyruvate pathway for IPP biosynthesis is an essential prerequisite to understanding and rationally manipulating terpenoid formation in plants.

Prenyltransferases and terpenoid synthases

In contrast to the confusion about IPP biosynthesis, the pathways and enzymes involved in the utilization of IPP in terpenoid biosynthesis are well-understood and widely accepted. The enzyme known as IPP isomerase catalyses the reversible conversion of IPP into DMAPP, which serves as the reactive allylic primer for elongation. A group of enzymes known collectively as prenyltransferases convert IPP and DMAPP into GPP, FPP and GGPP (Figure 1). These prenyl diphosphates serve as the immediate precursors for the synthesis of monoterpenes (from GPP), sesquiterpenes and triterpene-derived sterols (from FPP) and diterpenes and tetraterpene-derived carotenoids (from GGPP) (reviewed in [11,57]). Genes encoding FPP and GGPP synthases have been isolated from a number of plant sources (reviewed in [57,58]). The synthesis of FPP and FPP-derived sesquiterpenes and triterpenes is generally believed to occur in the cytoplasm, while the formation of GGPP and GGPP-derived diterpenes and tetraterpenes occurs in the plastid (reviewed in [57,59]). Although GPP synthase has been characterized and partially purified from a number of plant sources [60–62], the corresponding gene has not yet been cloned. GPP synthase has been localized in both the plastid of *Vitis vinifera* [63], where it is presumably involved in monoterpene biosynthesis, and in the cytoplasm of shikonin-producing cultures of *Lithospermum erythrorhizon* [64]. That monoterpene biosynthesis appears to be restricted to the plastid [52,65,66] argues for a plastidial localization of GPP synthase associated with essential oil formation. FPP synthase, in particular, has been extensively studied. As is the case with HMGR, recent work has demonstrated the presence of at least two genes for FPP synthase that are differentially expressed in different tissues of *Arabidopsis* [67]. The X-ray crystal structure of recombinant avian FPP synthase at 2.6 Å resolution has recently been reported [68]. Mechanistic studies using a variety of substrate analogues [69,70], and both site-directed [71–73] and random [74–76] mutagenesis of prenyltransferases, are providing valuable insights into the mechanism and product specificity of these enzymes, which are central to terpenoid biosynthesis.

A family of enzymes known collectively as terpenoid synthases are responsible for converting GPP, FPP and GGPP into the myriad array of monoterpenes, sesquiterpenes and diterpenes found in nature. Many of the products generated by these synthases are cyclic, and for this reason the enzymes are also often referred to as cyclases. Monoterpene synthases have been extensively studied in essential-oil-producing plants [57], whereas sesquiterpene synthases have been studied both in plants producing sesquiterpene-derived phytoalexins (see, for example, [77–79]) and in fungi producing sesquiterpene-derived toxins and antibiotics [80]. Diterpene synthases responsible for the formation of several natural products, such as the phytoalexin casbene [81,82], resin acids [83] and taxol [84], as well as *ent*-kaurene synthase [85–88] (involved in gibberellin biosynthesis) have been partially purified and characterized. These synthases have attracted attention not only because of their obvious importance as the enzymes responsible for the production of terpenoids, but also because of the unusual carbocationic reaction mechanisms involved in converting the acyclic, achiral allylic diphosphates into the wide variety of chiral cyclic structures found in nature. Detailed studies with several monoterpene and sesquiterpene synthases have elucidated the mechanism(s) of cyclization catalysed by these enzymes (reviewed in [57,80,89]). Only selected, recent examples of terpenoid synthases from plants that have been purified, characterized and cloned will be discussed here.

4(*S*)-(−)-Limonene synthase has been purified from spearmint (*Mentha spicata*) [90] and the liverwort *Ricciocarpos natans* [91] and characterized, and the corresponding cDNA has been isolated from both spearmint [92] and *Perilla frutescens* [93] and functionally expressed in *E. coli*. The cDNA for both spearmint and *Perilla* limonene synthases encodes a preprotein of approx. 60 kDa with a typical N-terminal plastidial transit peptide, confirming the localization of this enzyme in the plastid. In both cases, however, the precise cleavage site of this transit peptide has not yet been determined. An unusual feature of many terpenoid synthases is their ability to produce multiple products. This feature is considered to be a result of the highly reactive carbocationic intermediates generated at the enzyme active site, the variety of skeletal rearrangements these intermediates may undergo and the variety of ways in which the reaction can be quenched (reviewed in [57]). Recombinant limonene synthase from spearmint expressed in *E. coli* produces precisely the same spectrum of minor products observed with the native enzyme (Figure 4) thus providing conclusive evidence that the formation of these multiple products is not due to contaminating enzyme activities [92].

Sesquiterpene synthases of plant origin have been studied most commonly in tissues which accumulate sesquiterpene-derived phytoalexins. *epi*-Aristolochene synthase from tobacco (*Nicotiana tabacum*) was the first plant sesquiterpene synthase to be cloned and functionally expressed in *E. coli* [94]. More recently, vetispiradiene synthase and δ-cadinene synthase genes have been isolated from *Hyoscyamus muticus* [95] and cotton (*Gossypium arboreum*) [77], respectively. Vetispiradiene is believed to be the first committed precursor in the synthesis of the phytoalexin lubimin in several solanaceous species [79]. The cadinane family of sesquiterpenes are the most widely distributed sesquiterpenes found in plants [96]. A homology-based PCR cloning approach was employed to isolate a gene for δ-cadinene synthase, which is induced in response to fungal elicitation of tissue cultures of cotton. Close correlation of δ-cadinene synthase

Figure 4

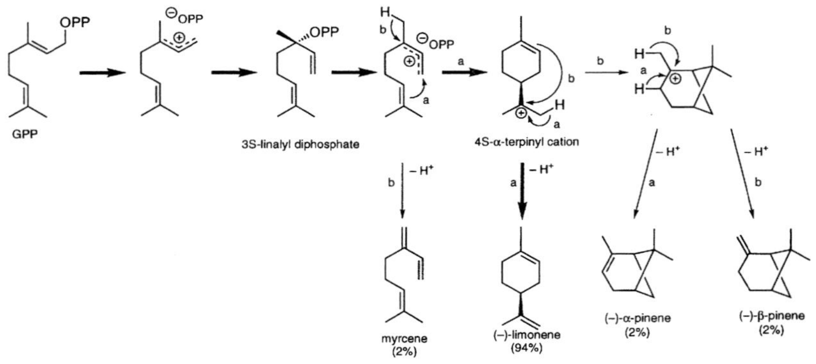

Formation of multiple products by 4(*S*)-(−)-limonene synthase

The major catalytic route from GPP to limonene is indicated with bold arrows

mRNA levels with induction of hemigossypol and gossypol production was observed, implicating δ-cadinene as a precursor of these phytoalexins [77].

Unlike the monoterpene and sesquiterpene synthases, diterpene synthases have evolved two different strategies for the cyclization of GGPP. The first strategy, as with monoterpene and sesquiterpene synthases, involves ionization of the diphosphate ester to initiate cyclization, whereas the second strategy involves cyclization initiated by protonation of the terminal double bond of GGPP. Casbene synthase [97] from castor bean (*Ricinus communis*) and taxadiene synthase [84] from Pacific yew (*Taxus brevifolia*) both catalyse cyclizations that are initiated by ionization of the diphosphate ester (Figure 5). Casbene is a diterpene phytoalexin produced in castor bean in response to fungal infection [98]. Taxadiene is the first committed intermediate in the biosynthesis of the chemotherapeutic agent taxol [99]. Taxadiene synthase has recently been partially purified from Pacific yew [84] and characterized, and a cDNA clone has been isolated and functionally expressed in *E. coli* [100]. A mechanistic study of taxadiene synthase has established the sequence of catalytic steps involved in the cyclization of GGPP to taxadiene (Figure 5) [101]. In contrast to casbene and taxadiene synthases, the initial step in the synthesis of *ent*-kaurene involves protonation of the terminal double bond of GGPP by *ent*-kaurene synthase A,

Figure 5

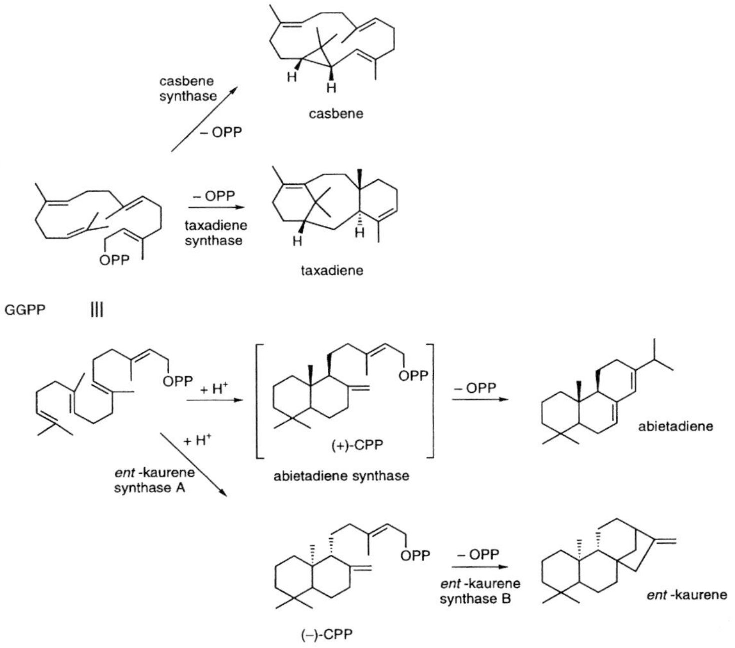

Two different strategies for the cyclization of GGPP

Casbene synthase and taxadiene synthase initiate cyclization by ionization of the diphosphate ester. Abietadiene synthase and ent-kaurene synthase A initiate cyclization by protonation of the terminal double bond.

initiating cyclization to (–)-copalyl diphosphate, which is subsequently cyclized to *ent*-kaurene by ionization of the diphosphate ester in a reaction catalysed by *ent*-kaurene synthase B (Figure 5) [86,88]. The genes for both *ent*-kaurene synthase A [85,87] and B [102] have recently been cloned.

Abietadiene is the first committed precursor in the biosynthesis of abietic acid [103,104], which is a major constituent in the rosin of pines, firs and spruces [105]. Abietadiene synthase has been purified from grand fir (*Abies grandis*) and characterized [106], and the cDNA has recently been isolated and functionally expressed in *E. coli* [83]. In contrast to the synthesis of *ent*-kaurene by two separate *ent*-kaurene synthases (A and B activities) abietadiene synthase catalyses both types of cyclization reactions [83]. Thus, cyclization to abietadiene is initiated by protonation of the terminal double bond of GGPP to produce enzyme-bound (+)-copalyl diphosphate, followed by ionization of the diphosphate ester of copalyl diphosphate to initiate the second cyclization, yielding abietadiene (Figure 5). It is interesting to note that abietadiene synthase shows non-overlapping regions of homology to both *ent*-kaurene synthase A and B, and it has been suggested that these regions correspond to domains of the enzyme responsible for the two separate cyclization reactions catalysed [83].

Comparison of the available cDNA sequences for the monoterpene, sesquiterpene and diterpene synthases and the prenyltransferases allows the identification of a number of common primary structural features of these proteins (reviewed in [57]). A common motif among the terpenoid synthases and prenyltransferases is the consensus sequence (I,L,V)XDDXX(D,E), which is believed to be involved in binding the Mg^{2+} (or Mn^{2+}) salt of the diphosphate moiety of the substrate [107]. Accordingly, prenyltransferases, which bind two diphosphate ester substrates (IPP and the allylic diphosphate), contain two such aspartic acid-rich regions, whereas terpenoid synthases, which bind only one such substrate, contain only one aspartate-rich region. The X-ray crystal structure of avian FPP synthase also supports the importance of these aspartic acid residues in binding the bivalent metal ion-complexed substrate [68]. The presence of histidine, cysteine and arginine residues at the active site of monoterpene synthases has been proposed, based on the sensitivity of these enzymes to chemical modification reagents (reviewed in [57]). This suggestion is supported by the observation of histidine, cysteine and arginine residues that are either highly conserved or strictly conserved in the deduced amino acid sequences of all synthases examined to this point. Plant monoterpene, sesquiterpene and diterpene synthases exhibit 30–40% identity and 50–60% similarity at the amino acid level, but they do not resemble closely the known fungal sesquiterpene synthases. The similarity of sequences of plant origin is proving to be valuable in the isolation and identification of new terpenoid synthases by homology-based PCR cloning approaches (J. Bohlmann and R. Croteau, unpublished work). Directed mutagenesis studies have been carried out with FPP synthase [71–73], and with fungal sesquiterpene synthases [108,109], demonstrating that it is possible to alter the products formed by these enzymes by modifying active site amino acid residues.

Summary

The field of plant terpenoid biosynthesis is developing rapidly on many fronts. However, there are significant areas that must be addressed before any rational attempts at bioengineering terpenoid formation can be attempted. The role of the GAP/pyruvate pathway in the formation of different families of isoprenoids, and the basic enzymology and regulation of this pathway all obviously need to be clarified. Otherwise, attempts to elucidate the control of the biosynthesis of specific terpenoids by focusing exclusively on the regulatory role of HMGR will be incomplete at best, or possibly completely wrong. The prospects for bioengineering terpenoid formation by modifying the activity of terpenoid synthases are considerably brighter at present. Our understanding of the mechanisms of both prenyltransferases and terpenoid synthases, and the availability of an increasing number of sequences for these enzymes, has set the stage for rationally manipulating the catalysts themselves and the pathways in which they participate.

References

1. Dawson, F.A. (1994) Naval Stores Rev. March/April, 6–12
2. Ohloff, G. (1994) Scent and Fragrances. Springer–Verlag, New York
3. Arbuck, S.G. and Blaylock, B.A. (1995) in Taxol: Science and Applications (Suffness, M., ed.), pp. 379–415, CRC Press, Boca Raton, FL
4. Holmes, F.A., Kudelka, A.P., Kavanagh, J.J., Huber, M.H., Ajani, J.A. and Valero, V. (1995) in Taxane Anticancer Agents: Basic Science and Current Status (Georg, G.I., Chen, T.T., Ojima, I. and Vyas, D.M., eds.), pp. 31–57, American Chemical Society, Washington D.C.
5. Woerdenbag, H.J., Pras, N., van Uden, W., Wallaart, T.E., Beekman, C. and Lugt, C.B. (1994) Pharm. World. Sci. **16**, 169–180
6. De Souza, N.J. (1993) J. Ethnopharmacol. **38**, 177–180
7. Flesch, V., Jacques, M., Cosson, L., Teng, B.P., Petiard, V. and Balz, J.P. (1992) Phytochemistry **31**, 1941–1945
8. Connolly, J.D. and Hill, R.A. (1992) Dictionary of Terpenoids, Chapman & Hall, New York
9. Gershenzon, J. (1994) J. Chem. Ecol. **20**, 1281–1328
10. Gershenzon, J. and Croteau, R. (1991) in Herbivores: Their Interaction with Secondary Metabolites (Rosenthal, G.A. and Berenbaum, M., eds.), pp. 165–219, Academic Press, New York
11. Gershenzon, J. and Croteau, R. (1993) in Lipid Metabolism in Plants (Moore, Jr., T.S., ed.), pp. 340–388, CRC Press, Boca Raton, FL
12. Bach, T.J. (1995) Lipids **30**, 191–202
13. Goldstein, J.L. and Brown, M.S. (1990) Nature (London) **343**, 425–430
14. Stermer, B.A., Bianchini, G.M. and Korth, K.L. (1994) J. Lipid Res. **35**, 1133–1140
15. Weissenborn, D.L., Denbow, C.J., Laine, M., Lång, S.S., Yang, Z., Yu, X. and Cramer, C.L. (1995) Physiol. Plant. **93**, 393–400
16. Campos, N. and Boronat, A. (1995) Plant Cell **7**, 2163–2174
17. Denbow, C.J., Lång, S. and Cramer, C.L. (1996) J. Biol. Chem. **271**, 9710–9715
18. Röhmer, M., Knani, M., Simonin, P., Sutter, B. and Sahm, H. (1993) Biochem. J. **295**, 517–524
19. Röhmer, M., Seeman, M., Horbach, S., Bringer-Meyer, S. and Sahm, H. (1996) J. Am. Chem. Soc. **118**, 2564–2566
20. Schwneder, J., Seeman, M., Lichtenthaler, H.K. and Röhmer, M. (1996) Biochem. J. **316**, 73–80
21. Schwarz, M.K. (1994) Terpen-Biosynthesis in *Gingko biloba*: Ein überraschende Geschichte., PhD thesis, Diss ETH No. 10951, Zürich
22. Cartayrade, A., Schwarz, M., Jaun, B. and Arigoni, D. (1994) Detection of two independent mechanistic pathways for the early steps of isoprenoid biosynthesis in *Ginkgo biloba*, poster presentation at 2nd Symposium of the European Network on Plant Terpenoids
23. Eisenreich, W., Menhard, B., Hylands, J.P., Zenk, M.H. and Bacher, A. (1996) Proc. Natl. Acad. Sci. U.S.A. **93**, 6431–6436
24. Nabeta, K., Ishikawa, T. and Okuyama, H. (1995) J. Chem. Soc., Perkin Trans. I, 3111–3115
25. Bach, T.J. and Lichtenthaler, H.K. (1987) in Ecology and Metabolism of Plant Lipids (Fuller, G. and Nes, W.D., eds.), pp. 109–139, American Chemical Society, Washington D.C.

26. Bach, T.J., Weber, T. and Motel, A. (1990) in Biochemistry of the Mevalonic Acid Pathway to Terpenoids. (Towers, G.H.N. and Stafford, H.A., eds.), pp. 1–82, Plenum Press, New York
27. Burden, R.S., Cooke, D.T. and Carter, G.A. (1989) Phytochemistry **28**, 1791–1804
28. Kim, K.K., Yamashita, H., Sawa, Y. and Shibata, H. (1996) Biosci. Biotechnol. Biochem. **60**, 685–686
29. Bach, T.J., Rogers, D.H. and Rudney, H. (1986) Eur. J. Biochem. **154**, 103–111
30. Heintze, A., Görlach, J., Leuschner, C., Hoppe, P., Hagelstein, P., Schulze-Siebert, D. and Schultz, G. (1990) Plant Physiol. **93**, 1121–1127
31. Hoppe, P., Heintze, A., Riedel, A., Creuzer, C. and Schultz, G. (1993) Planta **190**, 253–262
32. Heintze, A., Riedel, A., Aydogdu, S. and Schultz, G. (1994) Plant Physiol. Biochem. **32**, 791–797
33. Preiss, M. and Schultz, G. (1994) Plant Physiol. **105**, S65 (abstr. 318)
34. Hano, Y., Ayukawa, A., Nomura, T. and Ueda, S. (1994) J. Am. Chem. Soc. **116**, 4189–4193
35. Hano, Y., Ayukawa, A., Nomura, T. and Ueda, S. (1994) Naturwissenschaften **81**, 260–262
36. Hano, Y., Ayukawa, A. and Nomura, T. (1992) Naturwissenschaften **79**, 180–182
37. Hano, Y. and Nomura, T. (1995) Naturwissenschaften **82**, 376–378
38. Liedvogel, B. (1986) J. Plant Physiol. **124**, 211–222
39. Ohlrogge, J.B., Jaworski, J.G. and Post-Beittenmiller, D. (1993) in Lipid Metabolism in Plants (Moore, Jr., T.S., ed.), pp. 3–32, CRC Press, Boca Raton, FL
40. Roughan, P.G. and Ohlrogge, J.B. (1996) Plant Physiol. **110**, 1239–1247
41. Kennedy, I.A., Hill, R.E., Pauloski, R.M., Sawyer, B.G. and Spenser, I.D. (1995) J. Am. Chem. Soc. **117**, 1661–1662
42. Hill, R.E., Sawyer, B.G. and Spenser, I.D. (1989) J. Am. Chem. Soc. **111**, 1916–1917
43. White, R.L. and Spenser, I.D. (1982) J. Am. Chem. Soc. **104**, 4934–4943
44. Yokota, A. and Sasajima, K.I. (1986) Agric. Biol. Chem. **50**, 2517–2524
45. Riou, C., Tourte, Y., Lacroute, F. and Karst, F. (1994) Gene **148**, 293–297
46. Weber, T. and Bach, T.J. (1994) Biochim. Biophys. Acta **1211**, 85–96
47. Bach, T.J., Raudot, V., Vollack, K.-U., Weber, T. and Zeiler, S. (1994) Plant Physiol. Biochem. **32**, 775–783
48. Gondet, L., Weber, T., Maillot-Vernier, P., Benveniste, P. and Bach, T.J. (1992) Biochem. Biophys. Res. Commun. **186**, 888–893
49. Mackintosh, R.W., Davies, S.P., Clarke, P.R., Weekes, J., Gillespie, J.G., Gibb, B.J. and Hardie, D.G. (1992) Eur. J. Biochem. **209**, 923–931
50. Dale, S., Arró, M., Becerra, B., Morrice, N.G., Boronat, A., Hardie, D.G. and Ferrer, A. (1995) Eur. J. Biochem. **233**, 506–513
51. Ching, Y.P., Davies, S.P. and Hardie, D.G. (1996) Eur. J. Biochem. **237**, 800–808
52. McCaskill, D. and Croteau, R. (1995) Planta **197**, 49–56
53. McCaskill, D., Gershenzon, J. and Croteau, R. (1992) Planta **187**, 445–454
54. Wellburn, A.R. and Hampp, R. (1976) Biochem. J. **158**, 231–233
55. Schneider, M.M., Hampp, R. and Ziegler, H. (1977) Plant Physiol. **60**, 518–520
56. Soler, E., Clastre, M., Bantignies, B., Marigo, G. and Ambid, C. (1993) Planta **191**, 324–329
57. McCaskill, D. and Croteau, R. (1997) in Advances in Biochemical Engineering/Biotechnology (Scheper, T., ed.), pp. 107–146, Springer-Verlag, Berlin
58. Chen, A., Kroon, P.A. and Poulter, C.D. (1994) Protein Sci. **3**, 600–607
59. McGarvey, D.J. and Croteau, R. (1995) Plant Cell **7**, 1015–1026
60. Clastre, M., Bantignies, B., Feron, G., Soler, E. and Ambid, C. (1993) Plant Physiol. **102**, 205–211
61. Heide, L. and Berger, U. (1989) Arch. Biochem. Biophys. **273**, 331–338
62. Croteau, R. and Purkett, P.T. (1989) Arch. Biochem. Biophys. **271**, 524–535
63. Feron, G., Clastre, M. and Ambid, C. (1990) FEBS Lett. **271**, 236–238
64. Sommer, S., Severin, K., Camara, B. and Heide, L. (1995) Phytochemistry **38**, 623–627
65. Mettal, U., Boland, W., Beyer, P. and Kleinig, H. (1988) Eur. J. Biochem. **170**, 613–616
66. Pauly, G., Belingheri, L., Marpeau, A. and Gleizes, M. (1986) Plant Cell Rep. **5**, 19–22
67. Cunillera, N., Arró, M., Delourme, D., Karst, F., Boronat, A. and Ferrer, A. (1996) J. Biol. Chem. **271**, 7774–7780
68. Tarshis, L.C., Yan, M., Poulter, C.D. and Sacchettini, J.C. (1994) Biochemistry **33**, 10871–10877
69. Davisson, V.J. and Poulter, C.D. (1993) J. Am. Chem. Soc. **115**, 1245–1260
70. Davisson, V.J., Neal, T.R. and Poulter, C.D. (1993) J. Am. Chem. Soc. **115**, 1235–1245
71. Joly, A. and Edwards, P.A. (1993) J. Biol. Chem. **268**, 26983–26989
72. Koyama, T., Saito, K., Ogura, K., Obata, S. and Takeshita, A. (1993) Can. J. Chem. **72**, 75–79
73. Song, L. and Poulter, C.D. (1994) Proc. Natl. Acad. Sci. U.S.A. **91**, 3044–3048
74. Blanchard, L. and Karst, F. (1993) Gene **125**, 185–189

75. Ohnuma, S-i., Nakazawa, T., Hemmi, H., Hallberg, A.-M., Koyama, T., Ogura, K. and Nishino, T. (1996) J. Biol. Chem. **271**, 10087–10095
76. Ohnuma, S-i., Hirooka, K., Hemmi, H., Ishida, C., Ohto, C. and Nishino, T. (1996) J. Biol. Chem. **271**, 18831–18837
77. Chen, X.-Y., Chen, Y., Heinstein, P. and Davisson, V.J. (1995) Arch. Biochem. Biophys. **324**, 255–266
78. Threlfall, D.R. and Whitehead, I.M. (1988) Phytochemistry **27**, 2567–2580
79. Whitehead, I.M.M., Atkinson, A.L. and Threlfall, D.R. (1990) Planta **182**, 81–88
80. Cane, D.E. (1990) Chem. Rev. **90**, 1089–1103
81. Lois, A.F. and West, C.A. (1990) Arch. Biochem. Biophys. **276**, 270–277
82. Mau, C.J.D. and West, C.A. (1994) Proc. Natl. Acad. Sci. U.S.A. **91**, 8497–8501
83. Stofer Vogel, B., Wildung, M., Vogel, G. and Croteau, R. (1996) J. Biol. Chem. **271**, 23262–23268
84. Hezari, M., Lewis, N.G. and Croteau, R. (1995) Arch. Biochem. Biophys. **322**, 437–444
85. Benson, R.J., Johal, G.S., Crane, V.C., Tossberg, J.T., Schnable, P.S., Meeley, R.B. and Briggs, S.P. (1995) Plant Cell **7**, 75–84
86. Duncan, J.D. and West, C.A. (1981) Plant Physiol. **68**, 1128–1134
87. Sun, T. and Kamiya, Y. (1994) Plant Cell **6**, 1509–1518
88. Saito, T., Abe, H., Yamane, H., Sakurai, A., Murofushi, N., Takio, K., Takahishi, N. and Kamiya, Y. (1995) Plant Physiol. **109**, 1239–1245
89. Croteau, R. (1987) Chem. Rev. **87**, 929–954
90. Alonso, W.R., Rajaonarivony, J.I.M., Gershenzon, J. and Croteau, R. (1992) J. Biol. Chem. **267**, 7582–7587
91. Adam, K.-P., Crock, J. and Croteau, R. (1996) Arch. Biochem. Biophys. **332**, 352–356
92. Colby, S.M., Alonso, W.R., Katahira, E.J., McGarvey, D.J. and Croteau, R. (1993) J. Biol. Chem. **268**, 23016–23024
93. Yuba, A., Yazaki, K., Tabata, M., Honda, G. and Croteau, R. (1996) Arch. Biochem. Biophys. **332**, 280–287
94. Facchini, P.J. and Chappell, J. (1992) Proc. Natl. Acad. Sci. U.S.A. **89**, 11088–11092
95. Back, K. and Chappell, J. (1995) J. Biol. Chem. **270**, 7375–7381
96. Bordoloi, M., Shukla, V.S., Nath, S.C. and Sharma, R.P. (1989) Phytochemistry **28**, 2007–2037
97. Moesta, P. and West, C.A. (1985) Arch. Biochem. Biophys. **238**, 325–333
98. Sitton, D. and West, C.A. (1975) Phytochemistry **14**, 1921–1925
99. Koepp, A.E., Hezari, M., Zajicek, J., Stofer Vogel, B., LaFever, R.E., Lewis, N.G. and Croteau, R. (1995) J. Biol. Chem. **270**, 8686–8690
100. Wildung, M.R. and Croteau, R. (1996) J. Biol. Chem. **271**, 9201–9204
101. Lin, X., Hezari, M., Koepp, A.E., Floss, H.G. and Croteau, R. (1996) Biochemistry **35**, 2968–2977
102. Yamaguchi, S., Saito, T., Abe, H., Yamane, H., Murofushi, N. and Kamiya, Y. (1996) Plant J. **10**, 203–213
103. Funk, C., Lewinsohn, E., Stofer Vogel, B., Steele, C.L. and Croteau, R. (1994) Plant Physiol. **106**, 994–1005
104. Funk, C. and Croteau, R. (1994) Arch. Biochem. Biophys. **308**, 258–266
105. Lewinsohn, E., Savage, T.J., Gijzen, M. and Croteau, R. (1993) Phytochem. Anal. **4**, 220–225
106. LaFever, R.E., Stofer Vogel, B. and Croteau, R. (1994) Arch. Biochem. Biophys. **313**, 139–149
107. Ashby, M.N. and Edwards, P.A. (1990) J. Biol. Chem. **265**, 13157–13164
108. Cane, D.E., Shim, J.H., Xue, Q., Fitzsimmons, B.C. and Hohn, T.M. (1995) Biochemistry **34**, 2480–2488
109. Cane, D.E. and Xue, Q. (1996) J. Am. Chem. Soc. **118**, 1563–1564

The biorefinery: a quality investment opportunity

C. Kjøller
Bioraf Denmark Foundation, Research Centre, P.O. Box 35, DK-3720 Aakirkeby, Denmark

Abstract

The use of agricultural products in industry is not new. One hundred years ago, agriculture supplied a great deal of industry's raw materials, and even as late as the 1950s one-third of the cultivated area of W. Europe was used for raising crops for energy and industrial purposes. The advent of cheap oil eliminated most natural industrial products from the market.

The vision of the biorefinery concept is to strengthen the competitiveness of agriculture through the development of environmentally sound, quality products and production systems.

The concept takes an holistic viewpoint based on the market potential and the final potential of primary production, encouraging the development of environmentally and economically sustainable systems. Adapted and certified primary production creates the possibility of an optimized utilization of farm crops and thereby creates new sales outlets. The introduction of flexible technology provides the refinery with the possibility of economic adaptation to market changes in demand and price. The concept's production process aims at energy savings and at environmental and economic benefits based on optimized logistics through integrated and flexible production chains.

In general, agriculture lacks market alternatives and has made itself commercially vulnerable through a process whereby food and feed production have become the dominating factors in the Common Agricultural Policy of the European Union (E.U.).

It is therefore obvious and necessary that our present structure of cooperation between agriculture and industry should be revitalized and optimized to attain an improved utilization of agricultural raw-material biomass in industry.

Therefore we need to develop the biorefinery as 'the bridge between agriculture and industry'.

Introduction

There are many indications that society is on the threshold of developments whereby the greatest challenge is to preserve the quality of life through the development of products and services that are valuable for consumers as well as compatible with the environment in which we live. Based on a continually increasing interest in environmentally friendly manufacturing processes and goods, consumers have made it clear to manufacturers that attitudes are changing. Though this currently applies primarily to food, environmental constraints will undoubtedly be placed on other daily necessities as well. Another important development is the widespread awareness among the general public that many of the world's resources are finite. There is a great need for improved and responsible resource management.

Consumers are concerned about potentially harmful and allergenic chemical residues in products used daily, such as textiles, washing powders, packaging and cosmetics. People no longer wish to use products containing dangerous organic solvents such as paints and lacquers, wood preservatives and inks. Preferably, products should not pollute the environment after use but should be recyclable or degradable into substances that are harmless.

Politicians have acknowledged the need for changes in our production patterns by such policy statements as the Danish government's Environmental Policy Plan of Action, in which great emphasis is placed on phasing out substances harmful to the environment and on replacing non-renewable raw materials with renewable ones. A natural consequence of this new consciousness regarding our finite resources and of the acknowledgement that our world cannot in the long-run bear the pressure of increasing pollution and depletion will be an increased concentration on renewable raw materials.

We must rethink to manufacture products and develop processes that relieve the strain, instead of increasing it, on the world's finite resources. One such finite resource is crude oil, a resource that forms the foundation for our modern society. Not only do we burn oil, we also use it to produce a major part of our daily consumer goods such as plastics and textiles. In addition, the widespread use of fossil resources increases the content of atmospheric CO_2 year by year, and with it the risk of serious climatic changes. By increasing the use of biologically based products for energy and for the production of various consumer goods we will contribute to countering these trends. Renewable biological resources are CO_2 neutral, usually biodegradable and do not harm the environment. In addition, the processes for transforming the biological raw materials often require less energy and chemicals than corresponding petrochemical processes.

If we wish to accelerate developments by increasing our use of renewable raw materials, we must create a forum of awareness where the future is perceived as an holistic entity that emphasizes the interdependence of technology, the environment, production and utilization.

The biorefinery concept

Agriculture will undoubtedly experience these new trends, not only in demands for more environmentally friendly forms of production but also in new sales opportunities to industry. This will require, however, that farmers and farming organizations are prepared to produce raw materials for industry with the necessary quality and in sufficient amounts for the transition to a biologically-based, environmentally friendly and resource-economical industrial production.

The result will be a highly favourable atmosphere for close co-operation between agriculture and non-food industries. Not only will agriculture become a major supplier of raw materials to these industries, but the agricultural and food industries will also be major recipients of biologically based non-food products, such as foils and film wrapping for packaging of food and farming aids, geotextiles for soil improvement, polymers for encapsulating pesticides and seed grain, lubricants and bio-pesticides.

Farmers have experienced this situation previously, since farming and forestry have always produced raw materials for our daily necessities – not only food, but textiles, building materials, paper and cardboard and chemicals. One should not forget that, in the infancy of the industrial revolution, farming and forestry were industry's main source of raw materials, and the non-food sector was an important source of income for both farming and forestry. It is only within the past century that food production has assumed its current dominant role. Today a significant part of the agricultural production, and all of the production from forestry, is still used for non-food purposes. For example, half of the total production of starch and vegetable oil produced in the E.U. is used for non-food purposes. In this context it should be noted that an increase in non-food production does not necessarily mean a reduction in food production, since most farm crops contain components that can be used as raw materials for food *and* non-food production.

From 1991 to 1994, the 'Whole Crop Biorefinery Project' was carried out under the auspices of the E.U. ECLAIR (European collaborative linkage of agriculture and industry through research) programme. The project included the development and testing of technologies to fractionate, combine and upgrade crops, focusing on wheat and rape (Figure 1).

The following presents the biorefinery concept and some of the main project results with special emphasis on the refining and whole-crop utilization of wheat [1].

Biorefining is often associated with whole-crop harvesting. Whole-crop harvesting demands new storage methods for harvested crops to ensure year-round capacity utilization at the biorefinery. The concept also allows separate deliveries of seed and straw to the biorefinery, yet conventional harvesting technology involving combine harvesting and subsequent gathering and baling of the straw can still be employed. The same applies to the use of existing drying and storage facilities.

Organizationally speaking, the biorefinery bridges the gap between farmers and industry and manages the dual functions of crop processing and marketing. Since the individual farmer is unable to manage these tasks, the establishment of a biorefinery on a cooperative basis creates new utilization possibilities and markets.

Figure 1

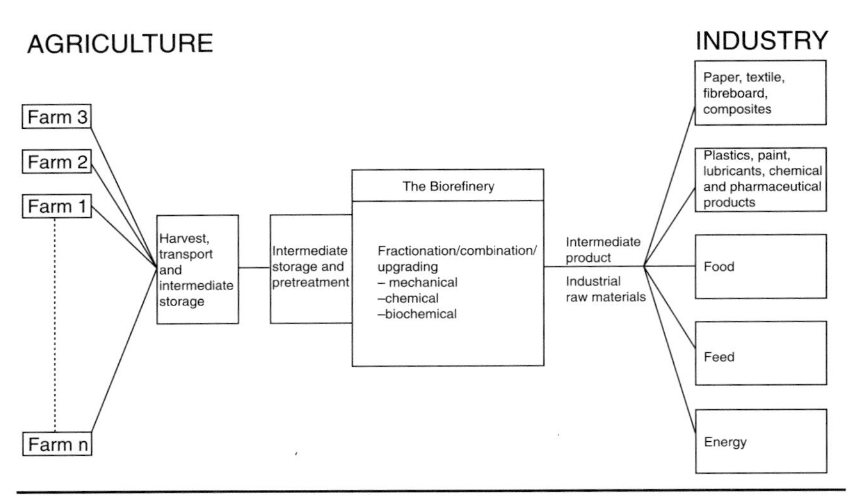

The Bioraf concept

Vision and strategic goals

In the time of our grandparents, farming was a major supplier to a number of industries such as the chemical, textile, and paint and lacquer industries. Today, agricultural raw materials are out-competed by cheap petrochemicals in most fields of industrial utilization. One of the prerequisites for an increased utilization of agricultural raw materials by industry is, therefore, the development of new processes and products.

Imagine, if you will, the optimization of utilization of the entire farm crop – the biomass. Could we, based on the utilization of biotechnology in its broadest sense (including plant breeding that incorporates genetic engineering), new harvesting, milling and separation techniques and on new organization, develop an improved and optimized production system; a more 'balanced production'? [2]. Although continued research and development are necessary they are in themselves not enough. It is of utmost importance that research and development continue in the next logical phase by demonstrating the new opportunities provided by the biorefinery to benefit agriculture and industry.

In general, agriculture lacks market alternatives and has made itself commercially vulnerable through a process whereby feed and food production have become the dominating factors in the E.U. Common Agriculture Policy. It is therefore obvious, but also necessary, that our present structure of co-operation between agriculture and industry be revitalized and optimized to attain an improved utilization of agricultural raw-material in industry. The aim, therefore, is to develop the biorefinery: a new, supplemental agriculture industry that in principle shall function in the same manner as an oil refinery. The biorefinery totally fractionates the raw materials into their chemical components, in accordance with market needs and profitability.

From idea to reality

The Bioraf Denmark Foundation is a non-profit making organization established in 1988 with the aim of co-ordinating Denmark's efforts to promote agri-industrial development and the new utilization of agricultural crops. Danish participation in the E.U. research and development programmes required international collaboration. The Bioraf Concept was developed and tested in the Whole Crop Biorefinery Project, involving the interdisciplinary co-operation of ten research institutions and industrial partners from five European countries [1]. The project lasted three and a half years from mid-1991 to the end of 1994, and consisted of six sub-projects. An overview of project activities and participants is shown in Figure 2.

A fundamental element of the project was the development and testing of technologies for the fractionation, combination and up-grading of crops and crop components. To serve these needs, an industrial pilot plant was built and full-scale machinery was installed. The pilot plant contained the following processing lines: (1) Dry straw fractionation. Cutting and chopping of straw, fractionation by means of a disk mill and separator into straw-chip fractions and straw meal. (2) A semi-wet straw process. Defibration of straw chips using a turbo-mill. (3) A dry wheat process: simplified disk milling process for manufacturing baking flour, industrial flour (suitable for starch production) and wheat bran. The ratio between industrial and baking flour can be varied. (4) A semi-wet wheat process: production of industrial flour and wheat bran using a turbo-mill and disk mill. Drying and grinding can be carried out simultaneously. (5) Wheat starch extraction. Industrial flour from the dry or semi-wet process is fractionated by decanters into wheat starch and gluten, as well as a wet residual product that can

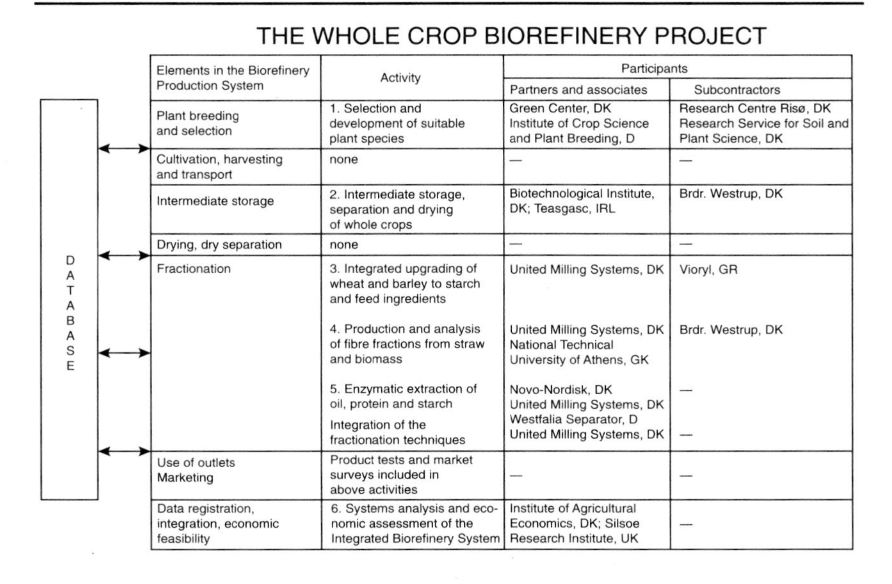

THE WHOLE CROP BIOREFINERY PROJECT

DATABASE

Elements in the Biorefinery Production System	Activity	Participants: Partners and associates	Participants: Subcontractors
Plant breeding and selection	1. Selection and development of suitable plant species	Green Center, DK Institute of Crop Science and Plant Breeding, D	Research Centre Risø, DK Research Service for Soil and Plant Science, DK
Cultivation, harvesting and transport	none	—	—
Intermediate storage	2. Intermediate storage, separation and drying of whole crops	Biotechnological Institute, DK; Teasgasc, IRL	Brdr. Westrup, DK
Drying, dry separation	none	—	—
Fractionation	3. Integrated upgrading of wheat and barley to starch and feed ingredients	United Milling Systems, DK	Vioryl, GR
	4. Production and analysis of fibre fractions from straw and biomass	United Milling Systems, DK National Technical University of Athens, GK	Brdr. Westrup, DK
	5. Enzymatic extraction of oil, protein and starch	Novo-Nordisk, DK United Milling Systems, DK Westfalia Separator, D	—
	Integration of the fractionation techniques	United Milling Systems, DK	—
Use of outlets Marketing	Product tests and market surveys included in above activities	—	—
Data registration, integration, economic feasibility	6. Systems analysis and economic assessment of the Integrated Biorefinery System	Institute of Agricultural Economics, DK; Silsoe Research Institute, UK	—

Figure 2

The Whole Crop Biorefinery Project

be used for feed. (6) Enzymic fractionation of rape. Pre-ground rapeseed undergoes enzymic digestion to break down cell walls. Subsequent separation using decanters and centrifuges gives oil, protein meal, syrup and hulls.

Alongside technological development and testing in the pilot plant, models of full-scale biorefinery systems were made. The models were used for systems analysis and economic evaluation. The main purpose of this sub-project was to provide an overview of the entire system and to determine whether crop fractionation provides sufficient added value to cover the refinement costs and compete with conventional crop processing and utilization. The commercial acceptance of biorefineries requires farmers to achieve higher crop prices and the resulting intermediate products to be sold to industry at competitive prices.

Bioraf models

Bioraf models based on whole-crop wheat and whole-crop oil-seed rape are presented in Figures 3 and 4, respectively. Though the models are simplified, they present a clear picture of the processes utilized and the mass balances based on inputs from 2500 ha, corresponding to 30000 metric tons of dry matter from whole-crop wheat and 21000 metric tons of dry matter from whole-crop oil-seed rape. While the dispersion ratios of the various fractions are guidelines, the ratios among the individual fractions can to a great extent be varied to meet market needs in actual situations.

As the mass balances show, the entire crop is utilised. This makes it possible to work with simplified and less expensive processing stages. For example, conventional starch processing requires the highest possible degree of extraction. However, since the crop component that is not used for starch is instead used for such products as baking flour, this opens up the possibility of reducing the number of milling steps and simplifying the entire process. The products obtained from whole-crop wheat using Bioraf model 1 are listed in Table 1.

Total utilization of the wheat plant requires links with a number of different markets and with purchasers of industrial non-food products, as well as with food, feed and energy producers. Similar products are obtained for whole-crop oil-seed rape using Bioraf model 2 (Table 2).

By combining models 1 and 2, and taking into account the utilization of other crops, it becomes clear that a broad range of industrial raw materials can be manufactured at a biorefinery.

Main results

The Bioraf Project succeeded in developing a technology that can be used for the fractionation, combination and upgrading of farm crops, initially wheat and oil-seed rape. Milling and starch separation processes were developed and tested in continuous operation, while enzymic fractionation remained a batch-based operation. Significant progress was achieved in regard to capacity utilization and simplification, reducing processing costs. It was also shown that plant breeding

Figure 3

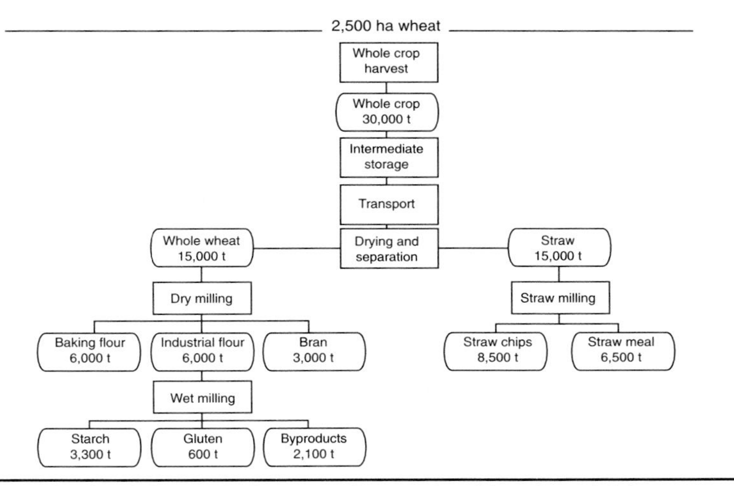

Products obtained from whole-crop wheat using Bioraf model 1

and genotype selection could be used to 'tailor' crops to the biorefinery and to new utilization purposes.

However, logistical problems concerned with the harvest, transport and, especially, storage of harvested whole crops were not fully solved. Based on our current knowledge, it is not economically or practically feasible to establish whole-crop harvesting in association with biorefineries to achieve one of the prerequisites for profitable operations: year-round capacity utilization of the plant. Biorefineries must presently, therefore, base their operations on conventional harvesting technologies (combine harvesting) and year-round deliveries of grain and straw from the farmers themselves. In other words, farmers must still bear the responsibility of managing the harvesting, drying and storage of agricultural produce.

Economic analyses show that, based on Danish conditions, good possibilities exist for profitable operations of a biorefinery similar to model 1

Figure 4

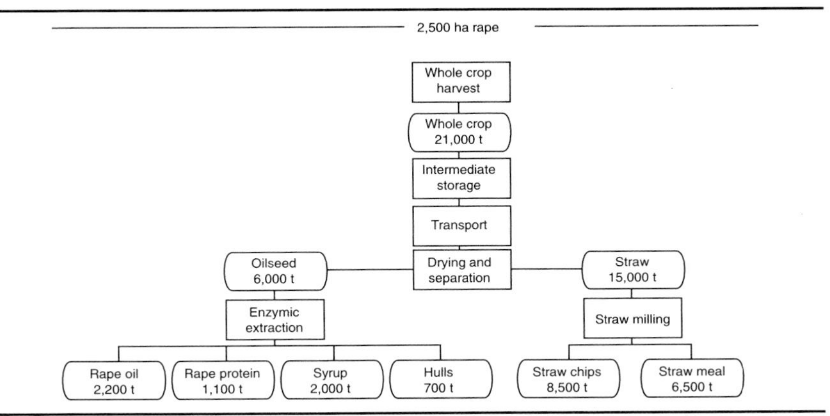

Products obtained from whole-crop oil-seed rape using Bioraf model 2

Table 1

Product	Further utilization
Straw chips	Composite industry
	Paper and pulp industry
	Packaging industry
	Energy
Straw meal	Energy
	Feed industry
Baking flour	Food industry
Bran	Feed industry
Starch	Food industry
	Paper industry
	Chemical and pharmaceutical industry
Gluten	Food industry
By-products	Feed, pig industry

Products obtained from whole-crop wheat using Bioraf model 1 (Figure 3)

Table 2

Product	Further utilization
Straw chips	Composite industry
	Paper and pulp industry
	Packaging industry
	Energy
Straw meal	Energy
Rape oil	Food industry
	Energy, bio-diesel
	Biodegradable lubricants
	Paint and lacquer industry
Rape protein	Feed industry
	Fermentation industry
	Food industry
Syrup	Feed industry
	Food industry
Hulls	Energy

Products obtained from whole-crop oil-seed rape using Bioraf model 2 (Figure 4)

(Figure 3) in which baking flour, starch, gluten, bran, by-products, straw chips and straw meal are produced from wheat grain and wheat straw. However, the enzymic fractionation of oil-seed rape (model 2, Figure 4) is not profitable, at least not on a scale relevant to a locally situated biorefinery. Processing costs require even greater reductions, and the potential for added values for oil and protein must be greater than those under current conditions.

Further development and perspectives

If biorefineries are to achieve a commercial breakthrough in increased utilization of farm crops for non-food uses, as well as for new food uses, additional work must be carried out that is based on the results of the Whole Crop Biorefinery Project. The technological developments achieved during the project have enabled effective raw material processing. The next phase will be the creation of market-related requirements for selling the raw materials, as well as the initial technological implementation.

The future strategies encompass three main elements:

Co-operation with receiver industries to carry out product development and test production

Work will be carried out on three areas: fibre, speciality flour/starch and oil.

In the field of fibre, efforts will focus on: (1) developing processes for defibration, drying and separation; (2) developing fibre mats for form-fitted products and insulation materials; and (3) 'tailoring' fibres based on plant breeding and selection to simplify the defibration process

In the field of speciality flour/starch, efforts will focus on: (1) continued development/integration of milling and starch production processes aimed at the manufacture of speciality flour, modified starch and hydrolysed fibres/by-products; (2) quality control, including environmentally sound or organic production; and (3) 'tailoring' crops through plant breeding and selection

In the field of rapeseed oil and enzyme processes, efforts will focus on: (1) developing oil for new lubricants and 'tailoring' special oils for specific purposes; (2) developing high quality edible oils; (3) developing protein of high biological value; (4) combining feed fractions; and (5) removing glucosinolates from the syrup fraction and developing biodegradable bio-pesticides

The establishment of first-generation, full-scale demonstration plants based on the most promising technology

Future efforts will focus on constructing a full-scale plant for the biorefining of wheat (model 1, Figure 3), which is the most promising option both from technical and economic viewpoints. The main purposes of a demonstration plant will be to show that profitable operations are possible and to acquire experience to facilitate implementation.

Continued system analyses, market studies and economic evaluation

It is necessary to follow up these activities with continued system analysis, market studies and economic evaluation, thereby creating an overview of the entire system from farmer to biorefinery to receiver industries to end users. Bottlenecks, as well as the most interesting opportunities, shall be identified and the commercial potential evaluated. The driving forces for biomass-based chemistry are summarized in Figure 5.

Conclusions

A rapid commercial breakthrough for increased non-food use of farm crops is not 'just around the corner'. Promising possibilities exist in certain areas, but it is difficult to compete with existing technologies and raw materials. Raw materials based on fossil fuels are relatively inexpensive and the technology to exploit fossil fuels is well-developed and widespread following many years of utilization.

For almost a generation we have surrounded ourselves with increasing numbers of synthetic products. Now we are experiencing the occurrence of new, but not always totally clear, negative correlations between the synthetic products with which we surround ourselves and our health and environment. There are already many examples of products that originally were considered harmless but that were subsequently shown to have negative effects on our health and/or environment [such as organic solvents, asbestos fibres, plasticizers and certain synthetic surface-active ingredients (soaps)].

Figure 5

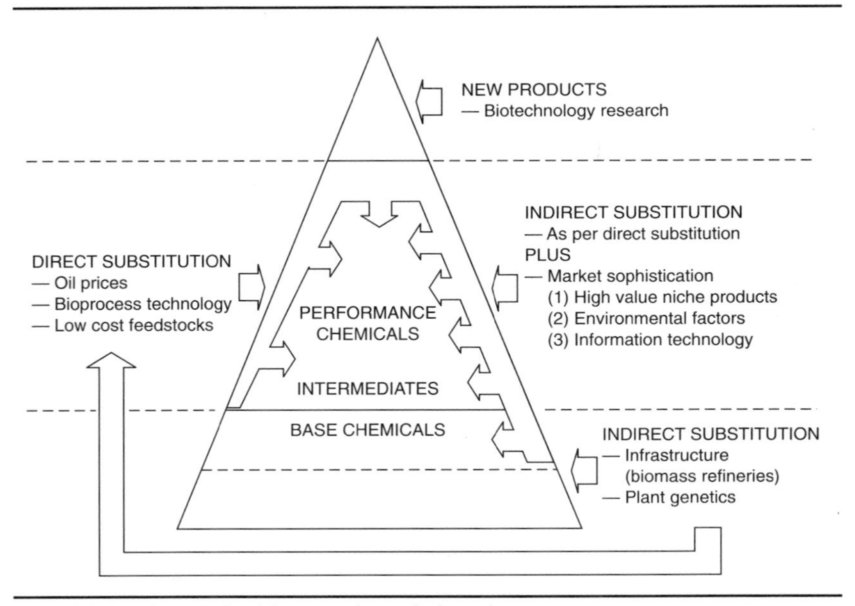

The driving forces for biomass-based chemistry

The business and industrial communities are aware of these issues and there is today a clear, but still gradual, trend away from synthetic towards 'natural' products, that is, products based on biological raw materials. For example: (1) natural fibres are today being used to some extent to replace asbestos fibres; (2) automobile manufacturers such as Mercedes, Volkswagen and Fiat use natural fibres to an increasing extent as inside coverings in doors, on the instrument panel, in ceiling covers etc.; (3) the electronics industry is experimenting with the utilization of biopolymers for manufacturing print plates and cabinets; (4) the paint and lacquer industry, the printing ink industry, the wood preservative industry and others have increased their use of products based on biological raw materials, for example, as substitutes for fossil-fuel solvents; (5) the detergent industry is substituting synthetic surfactants with surfactants made of biological raw materials; and (6) the packaging industry, the food industry and retailers are experimenting with environmentally friendly packaging, shopping bags etc.

Finding new uses and new markets for European agricultural production has been discussed for decades in the E.U. and there have been notable successes in the food and fibre areas. The variety of choices in fruits, vegetables, meats, dairy products, cereals and bakery goods is overwhelming. However, the opportunity for market growth in these product areas is very limited. People can only eat so much food.

The new opportunity for market growth is now to be found in the non-food industrial area as discussed above. With good science, imagination and access to capital, a thriving renewable base industry could provide the market expansion needed for an economically healthy agricultural industry. Biorefineries offer the opportunity to combine environmental and economic development objectives to generate many hightech/highpay jobs in rural communites of the E.U., where under-utilized infrastructures would welcome employment opportunities. However, the problem in the E.U. seems to be the lack of a 'market pull' policy instead of a 'technology push' policy. Part of the answer is that there has not been consistent agricultural impetus to realize the full potential for new industrial uses and higher priority has been placed on preserving E.U. commodity support programmes and export restitutions.

The World Trade Organisation emphasizes that the long-term objective is to attain substantial and gradual reductions both in subsidies and in the protection of agriculture, which shall lead to an adjustment of prices to avoid restrictions and turbulence on world markets for agricultural products.

The World Trade Agreement on Agriculture, together with the certainty of reduced E.U. support for agriculture, mean that farm organizations and the European agro-industry need to re-examine non-food industrial markets and promote an interaction with new start-up entrepreneurs, emphasizing market pull support rather than technology push. Commitment must be consistent and should include a transitional involvement of the public sector. A part of the current E.U. Common Agricultural Policy budget could be moved to expand forthcoming Framework Research Programmes and to support joint private/public investment opportunities in the member countries.

Unless E.U. farm organizations and the agro-industry pursue this option a considerable part of the Common Agricultural Policy budget for supporting agricultural production could be removed without an alternative system being put in place.

The most important step is to interest industry in actively participating in the development of the products and markets. This may be difficult since economic incentives are questionable. Therefore, a breakthrough also requires an improved climate of progress that should include public acceptance (sound regulations and clear information), industrial commitment (non-discriminatory regulations and intellectual property protection) and shared rewards (by industry, the farmer and the consumer).

References

1. Gylling, M. (1995) The Whole Crop Biorefinery Project, Main Report The Bioraf Denmark Foundation, October 1995, pp. 25–29, The Bioraf Denmark Foundation, Aakirkeby, Denmark
2. Rexen, F. and Munck, L. (1984) Cereal Crops for Industrial Use in Europe, Carlsberg Research Laboratory, Copenhagen, June 1984, E.U. Commission, Luxembourg